RAND PROJECT AIR FORCE

T0302550

Defeat, Not Merely Compete

China's View of Its Military Aerospace Goals and Requirements in Relation to the United States

Scott W. Harold

Prepared for the United States Air Force
Approved for public release; distribution unlimited

For more information on this publication, visit www.rand.org/t/RR2588

Library of Congress Cataloging-in-Publication Data is available for this publication.

ISBN: 978-1-9774-0100-7

Published by the RAND Corporation, Santa Monica, Calif.

Cover: caption/source information.

Preface

As China develops its armed forces, what role does competition with the United States play in shaping the military aerospace capabilities development of the People's Liberation Army (PLA)? The U.S. Air Force (USAF) vision statement lays out a goal of "global vigilance, global reach, and global power," a set of ambitions that PLA authors appear to be mimicking with their "bright eyes, strong fists, and long arms" (光眼，重拳，长臂) slogan for the PLA Air Force (PLAAF). Similarly, Chinese authors treat the quest for the capacity to "simultaneously be able to conduct offensive and defensive integrated air and space operations" in ways that appear to mimic the thinking of the U.S. armed forces on military aerospace. This study examines how the PLA, in seeking to accomplish the Chinese Communist Party's missions, strives to match or exceed the capabilities of the United States in military aerospace. It also explores how, in benchmarking its ambitions against the U.S. military, the PLA approaches the question of whether to copy from a leading foreign aerospace power or to develop a new and innovative approach to accomplishing a mission or fielding a capability. The study reviews the drivers for Chinese military aerospace development and identifies instances when China has copied or innovated in military aerospace development over the past two decades while also noting those areas where China has chosen not to compete.

The research reported here was commissioned by the USAF and conducted within the Strategy and Doctrine Program of RAND Project AIR FORCE as part of a fiscal year 2017 project focused on the growing reach of Chinese aerospace capabilities.

RAND Project AIR FORCE

RAND Project AIR FORCE (PAF), a division of the RAND Corporation, is the U.S. Air Force's federally funded research and development center for studies and analyses. PAF provides the Air Force with independent analyses of policy alternatives affecting the development, employment, combat readiness, and support of current and future air, space, and cyber forces. Research is conducted in four programs: Force Modernization and Employment; Manpower, Personnel, and Training; Resource Management; and Strategy and Doctrine. The research reported here was prepared under contract FA7014-16-D-1000.

Additional information about PAF is available on our website:
www.rand.org/paf/

This report documents work originally shared with the U.S. Air Force in September 2017. The draft report, issued on September 26, 2017, was reviewed by formal peer reviewers and U.S. Air Force subject-matter experts.

Contents

Preface ... iii
Summary ... vii
Acknowledgments ... xi
Abbreviations ... xiii
1. Introduction ... 1
Organization of the Report ... 4
2. Terminology, Methodology, and Data Sources ... 7
3. Rapid Recent Development from a Low Post–Cold War Baseline ... 11
4. PLA Aims to Defeat, Not Merely Compete with, U.S. Military ... 17
5. When Does China Copy Foreign Militaries? ... 23
6. China's Innovations in Military Aerospace ... 29
7. Conclusions and Implications ... 35
Appendix A. Transfers of Air Defense Systems, China to World (Final Deliveries and New Orders, 2010 to 2017) ... 39
Appendix B: Transfers of Combat Aircraft, China to World (Final Deliveries and New Orders, 2010 to 2017) ... 41
Appendix C: Transfers of Missiles, China to World (Final Deliveries and New Orders, 2010 to 2017) ... 45
References ... 51

Summary

Over the past two decades, the Chinese People's Liberation Army (PLA) has made rapid advances in building up new capabilities and operational concepts. Aerospace power has been a core feature of the PLA's rapid modernization. In particular, since 2004, the PLA Air Force (PLAAF) has pursued a service strategy aimed at developing the capacity to "simultaneously prosecute offensive and defensive integrated air and space operations" (空天一体，攻防兼备). During this period, Chinese military authors have written about transforming the PLAAF into what they refer to as a "strategic air force," one that can "move beyond its [traditional] focus on air defense of Chinese territory" and "directly support national policy objectives and achieve a wide range of strategic goals."[1] One recent study of Chinese military aerospace writings found that many PLA authors specifically cite the U.S. Air Force (USAF) as "a model for at least some aspects" of the PLAAF's transformation into a strategic air force.[2] In 2013, the USAF released a vision statement, titled *Global Vigilance, Global Reach, Global Power for America*, that described its roles and ambitions;[3] the PLA's views on importance of integrated air and space operations are similar to those outlined in this vision statement, and the PLA wants to support goals strikingly similar to the USAF's concepts of "vigilance," "reach," and "power." In addition, in 2014, the chief of staff of the USAF argued that achieving USAF core missions required "integrating airpower capabilities within and across air, space, and cyberspace."[4] The Chinese military appears to be prioritizing similar developments in the arenas of strategic intelligence, surveillance, and reconnaissance (ISR); tactical and strategic lift; and strike assets.

This report explores the extent to which the desire to "compete"[5] with the USAF (or other advanced air forces) shapes PLA thinking about the development of military aerospace power. I

---

[1] Michael S. Chase and Cristina L. Garafola, "China's Search for a Strategic Air Force," *Journal of Strategic Studies,* Vol. 39, No. 1, 2015, pp. 1–26.

[2] Michael S. Chase, Cristina L. Garafola, and Nathan Beauchamp-Mustafaga, "Chinese Perceptions of and Responses to US Conventional Military Power," *Asian Security*, 2017, pp. 1–19.

[3] Headquarters, Department of the Air Force, *Global Vigilance, Global Reach, Global Power for America*, Washington, D.C.: 2013.

[4] Mark A. Welsh, III, "Global Vigilance, Global Reach, Global Power for America: The World's Greatest Air Force—Powered by Airmen, Fueled by Innovation," *Air and Space Power Journal*, Vol. 28, No. 2, March–April 2014, pp. 4–10.

[5] I use the term *compete* to refer to a deliberate effort to develop capabilities designed to keep up with, match, and (if possible) exceed a real or potential rival.

examine how China selects between the options of *copying*[6] foreign powers and *innovating*[7] its own solutions to various operational military problems, as well as which areas China chooses to not compete in at all.

The Chinese military's decisions about whether and how to compete and when to copy or innovate can have serious consequences in terms of how the PLA, and especially those branches that operate some of China's most consequential power projection capabilities, are organized and operate. Recently, China has undertaken substantial structural reforms, many of which remade command and control relationships in ways that some leading PLA watchers assess as being inspired at least partly by the U.S. Goldwater-Nichols Act of 1986.[8] The reforms transformed former military regions into theater commands, which are now intended to receive and command forces provided by PLA service branches. In their new roles, the PLA Army, PLAAF, and PLA Navy (PLAN) are intended to function more like force providers, playing a role similar to the "organize, train, and equip" function that defines the relationship between the U.S. military services and the unified combatant command structure.[9] While it is unclear exactly how the PLARF (a service branch, but not clearly acting as a force provider to local theater commanders) and PLASSF (not a service branch, but an inter-service military organization, together with the PLA Joint Logistics Support Force) will be integrated under this new arrangement, these forces do represent China's recognition of the importance of aerospace power elements other than fixed-wing and rotary-wing airframes.[10] Many U.S. observers have also seen similarities in PLA hardware and operating practices that, at least superficially, appear to be derived from the observation or theft of the United States' or other countries' technologies and practices.[11] For

---

[6] *Copying* refers to the process of stealing or imitating foreign technology designs or operating patterns and lessons, even if some adaptation of the hardware designs or operational behavior occurs after adoption.

[7] By *innovating*, we mean developing a new technology substantially through one's own efforts, or solving a military problem or task in a new or novel way through operational or organizational responses that were generated indigenously.

[8] James Mulvenon, "China's 'Goldwater-Nichols': The Long-Awaited PLA Reorganization Has Finally Arrived," *China Leadership Monitor*, No. 49, pp. 1–6; Phillip C. Saunders and Joel Wuthnow, "China's Goldwater–Nichols? Assessing PLA Organizational Reforms," *Joint Forces Quarterly*, Vol. 82, July 2016, pp. 68–75.

[9] Joel Wuthnow and Phillip C. Saunders, *Chinese Military Reforms in the Era of Xi Jinping: Drivers, Challenges, and Implications*, Washington, D.C.: National Defense University, 2017. One important difference between the PLA and the U.S. military is that we do not know the extent to which the strategic assets of the PLAN (submarines equipped with nuclear weapons), the PLA Rocket Force (PLARF; it operates China's intercontinental ballistic missiles and nuclear weapons), and the PLA Strategic Support Force (PLASSF; it appears to control space and cyberspace assets) will be controlled by the theater commanders. It appears that control over these capabilities is likely to remain in the hands of the Central Military Commission.

[10] Cristina L. Garafola, "Will the PLA Reforms Succeed?" *China Analysis—European Council on Foreign Relations*, March 30, 2016.

[11] Analysts have pointed to a variety of at least superficial similarities between Chinese aerospace hardware and operational behavior and those of foreign predecessors. These have included aspects of the J-10, J-11, J-16, J-20, and J-31 fighters; the design of various Chinese unmanned aerial vehicles, including the Wing Loong, Caihong, and Xiang Long platforms; and the dress and operational behavior of Chinese aircraft carrier landing deck crews. See Brendan McGarry, "China's Fighters, Drone Look Like U.S. Aircraft," *DefenseTech*, June 20, 2013; Richard

these reasons, it is important to understand how China's military aerospace power is developing; when and why the PLA competes with the USAF on military aerospace; and how the Chinese leadership decides whether to copy or innovate in seeking to achieve these goals.

This report concludes that the PLA seeks to compete with the U.S. military not as a goal in and of itself, but rather as a means to achieving the political goals that the Chinese Communist Party (CCP) sets for the PLA—goals defined by the CCP's threat perceptions and policy ambitions. More specifically, the PLA seeks not merely to compete with, but to defeat, the U.S. military, should the two countries ever come into direct confrontation; the overwhelming majority of China's military capabilities developments and reforms, including its military aerospace capabilities developments, have been oriented toward this goal.[12] It is important to recognize that many of the PLA's efforts in the military aerospace sector focus on fielding specific capabilities in sufficient quantities to deter the United States from entering a conflict; the PLA would vastly prefer this over victory through combat.

In the quest to accomplish this, copying as well as innovating are valid pathways, although urgency and low costs appear to have made copying the preferred approach. However, the PLA copies only aspects of U.S. or foreign practices and capabilities relevant to China's goals and adapts them to local realities.

For the USAF, these findings offer some insights into the drivers behind the choices China makes to invest in specific domains related to ISR, strategic and tactical lift, and strike platforms and assets, as well as power projection in and through space and against space-based satellite architectures. A deeper knowledge of China's military aerospace strengths and weaknesses can help steer the USAF away from areas in which China has robust capacity and toward areas where its weaknesses could be better exploited. Analyses of Chinese aerospace development can also help the USAF (and the U.S. military and national security establishment more broadly) to identify possible or probable vectors along which the PLA is likely to target U.S. interests in any armed clash. Such knowledge can also be used to avoid certain types of military-to-military contact that might unintentionally help the PLA further advance its capabilities. In addition, a deeper understanding of when Chinese military aerospace development copies the United States or other countries' military aerospace power could help more effectively target U.S. denial and deception efforts.

Whittle, "New China Drone: Looks Like a Reaper, But . . . ," *Breaking Defense*, September 2, 2015; David Axe, "Go Ahead China—Copy Our Crappiest Warplane," *Daily Beast*, June 5, 2015; Sebastien Roblin, "China Stole This Fighter from Russia—and It's Coming to the South China Sea," *National Interest*, July 24, 2016; Bill Gertz, "Top Gun Takeover: Stolen F-35 Secrets Showing Up in China's Stealth Fighter," *Washington Free Beacon*, March 13, 2014; Andrew Erickson and Gabe Collins, "Chinese Aircraft Carrier Style! Assessing the First Takeoff and Landing," *Wall Street Journal*, November 27, 2012; Marcus Weisgerber, "China's Copycat Jet Raises Questions About F-35," *DefenseOne*, September 23, 2015.

[12] This assessment of the importance of the United States as the main military competitor and pacing challenge that the PLA plans against is derived from the interviews done for this study, as well as from Chase, Garafola, and Beauchamp-Mustafaga, 2017, and other sources consulted in the course of this study.

Acknowledgments

The author wishes to thank the following individuals for agreeing to share their insights during the course of the research for this study: Ken Allen, Dean Cheng, Tai Ming Cheung, Roger Cliff, Ian Easton, Richard Fisher, Kevin Pollpeter, Phillip C. Saunders, Mark Stokes, Larry Wortzel, Joel Wuthnow, and a small number of U.S. government analysts who must remain anonymous. The author is also grateful to Michael Chase for helpful guidance and feedback throughout the research process, and to Astrid Cevallos and Arthur Chan for research assistance. During the review process, Paula Thornhill, Cortez A. Cooper, III, and Michael Lostumbo provided useful feedback and suggestions for improving the paper.

Abbreviations

ASCM	antiship cruise missile
AWACS	airborne warning and control system
C2	command and control
CAS	close air support
CCP	Chinese Communist Party
CMC	Central Military Commission
COSTIND	Committee on Science and Technology in National Defense
FGA	fighter/ground aircraft
GAD	General Armaments Department
ISR	intelligence, surveillance, and reconnaissance
PAF	Project AIR FORCE
PLA	People's Liberation Army
PLAA	PLA Army
PLAAF	PLA Air Force
PLAN	PLA Navy
PLANAF	PLA Navy Air Force
PLARF	PLA Rocket Force
PLASSF	PLA Strategic Support Force
SAM	surface-to-air missile
UAV	unmanned aerial vehicle
UCAV	unmanned combat aerial vehicle
USAF	U.S. Air Force

1. Introduction

Over the past two decades, the Chinese People's Liberation Army (PLA) has rapidly advanced its new capabilities and operational concepts. Aerospace power has been at the center of this rapid modernization; since 2004, the PLA Air Force (PLAAF) has pursued a service strategy aimed at developing the capacity to "simultaneously prosecute offensive and defensive integrated air and space operations" (空天一体，攻防兼备). During this period, Chinese military authors have written about transforming the PLAAF into what they refer to as a "strategic air force," one that can "move beyond its [traditional] focus on air defense of Chinese territory" and "directly support national policy objectives and achieve a wide range of strategic goals."[1]

One recent study of Chinese military aerospace writings found that many PLA authors specifically cite the U.S. Air Force (USAF) as "a model for at least some aspects" of the PLAAF's transformation into a strategic air force.[2] In 2013, the USAF released a vision statement describing its roles and ambitions, titled *Global Vigilance, Global Reach, Global Power for America.*[3] Then–chief of staff of the USAF General Mark A. Welsh, III, subsequently argued that achieving USAF core missions requires "integrating airpower capabilities within and across air, space, and cyberspace."[4] The Chinese military appears to be prioritizing similar development of strategic intelligence, surveillance, and reconnaissance (ISR); tactical and strategic lift; and strike assets in support of goals that bear a striking resemblance to the USAF's concepts of vigilance, reach, and power. In addition, the PLA's view of the importance of integrated air and space operations corresponds to the views outlined in the USAF vision statement.

To what extent does a desire to compete with the USAF (or other advanced air forces) shape the PLA's thinking about the development of military aerospace power?[5] How does China select

---

[1] Michael S. Chase and Cristina L. Garafola, "China's Search for a Strategic Air Force," *Journal of Strategic Studies*, Vol. 39, No. 1, 2015, pp. 1 26.

[2] Michael S. Chase, Cristina L. Garafola, and Nathan Beauchamp-Mustafaga, "Chinese Perceptions of and Responses to US Conventional Military Power," *Asian Security*, 2017, pp. 1–19.

[3] Headquarters, Department of the Air Force, *Global Vigilance, Global Reach, Global Power for America*, Washington, D.C., 2013.

[4] Mark A. Welsh, III, "Global Vigilance, Global Reach, Global Power for America: The World's Greatest Air Force—Powered by Airmen, Fueled by Innovation," *Air and Space Power Journal*, Vol. 28, No. 2, March–April 2014, pp. 4–10.

[5] I use the term *compete* to refer to a deliberate effort to develop capabilities designed to keep up with, match, and (if possible) exceed a real or potential rival.

between copying what foreign powers do and innovating its own solutions to various operational military problems?[6] And which areas does China choose not to compete in at all?

The Chinese military's decisions about whether and how to compete, copy, or innovate can have serious consequences in terms of how the PLA, and especially those branches that operate some of China's most consequential power projection capabilities, are organized, trained, and equipped. For example, in recent years China has undertaken substantial structural reforms, many of which retooled command and control (C2) relationships in ways that some leading PLA watchers assess as being inspired at least in part by the U.S. Goldwater-Nichols Act of 1986.[7] The PLA's reforms have transformed the former military regions into theater commands, which are now intended to receive and command forces provided by the PLA's service branches. In their new roles, the PLA Army (PLAA), PLAAF, and PLA Navy (PLAN) are intended to function more like force providers, playing a role similar to the "organize, train, and equip" function that defines the relationship between the U.S. military services and the unified combatant command structure.[8] While it is not as clear exactly how the PLARF (a service that currently does not clearly act as a force provider to local theater commanders) and the PLASSF (not a service but an interservice military organization, together with the PLA Joint Logistics Support Force) will be integrated under this new arrangement, these forces do represent a recognition of the importance of elements of aerospace power other than just fixed- and rotary-wing airframes.[9] Many U.S. observers have also seen similarities in PLA hardware and operating practices that, at least superficially, appear to be derived from the observation or theft of the United States or other countries' technologies and practices.[10] For these reasons, it is important

---

[6] *Copying* refers to the process of stealing or imitating foreign technology designs or operating patterns and lessons, even if some adaptation of the hardware designs or operational behavior occurs after adoption. However, the nontechnical, open-source materials used as the sources for this report cannot be used to definitively state that Chinese hardware is a direct copy of U.S. platforms or that China's operational behavior or doctrine copies the U.S. military or another military. I define *innovating* as developing a new technology substantially through one's own efforts or solving a military problem or task in a new way through indigenous operational or organizational responses.

[7] James Mulvenon, "China's 'Goldwater-Nichols': The Long-Awaited PLA Reorganization Has Finally Arrived," *China Leadership Monitor*, No. 49, pp. 1–6; Phillip C. Saunders and Joel Wuthnow, "China's Goldwater–Nichols? Assessing PLA Organizational Reforms," *Joint Forces Quarterly*, Vol. 82, July 2016, pp. 68–75.

[8] Joel Wuthnow and Phillip C. Saunders, *Chinese Military Reforms in the Era of Xi Jinping: Drivers, Challenges, and Implications*, Washington, D.C.: National Defense University, 2017. One important difference between the PLA and the U.S. military is that we do not know the extent to which the strategic assets of the PLAN (submarines equipped with nuclear weapons), the PLA Rocket Force (PLARF; it operates China's intercontinental ballistic missiles and nuclear weapons), and the PLA Strategic Support Force (PLASSF; it appears to control space and cyberspace assets) will be controlled by the theater commanders. It appears that control over these capabilities is likely to remain in the hands of the Central Military Commission (CMC).

[9] Cristina L. Garafola, "Will the PLA Reforms Succeed?" *China Analysis—European Council on Foreign Relations*, March 30, 2016.

[10] Analysts have pointed to a variety of at least superficial similarities between Chinese aerospace hardware and operational behavior and those of foreign predecessors. These have included aspects of the J-10, J-11, J-16, J-20, and J-31 fighters; the design of various Chinese unmanned aerial vehicles (UAVs), including the Wing Loong,

to understand how China's military aerospace power is developing, when and why the PLA competes with the USAF on military aerospace, and how the Chinese leadership decides whether to copy or innovate in seeking to achieve these goals.

This study concludes that the PLA seeks to compete with the U.S. military not as a goal, but rather as a means to achieving the political goals set before the PLA by the Chinese Communist Party (CCP), goals defined by the CCP's threat perceptions and policy ambitions. More specifically, the PLA seeks not merely to compete with, but to deter and, if necessary, defeat, the U.S. military should the two countries ever come into direct confrontation; the overwhelming majority of China's military capabilities developments and reforms, including its military aerospace capabilities developments, have been oriented toward this goal.[11]

This has led China to prioritize the acquisition of advanced air superiority fighters; ballistic and cruise missiles; advanced integrated air defenses anchored on a network of surface-to-air missiles (SAMs); airborne and space-based ISR platforms; airborne C2 assets; space and counterspace assets; and, to a more limited extent, heavy lift capabilities. By contrast, the PLA has placed less (though growing) emphasis on aerial refueling and deck-based fixed-wing aircraft, has focused only belatedly on developing a new generation of manned bombers, and has put very little focus on close air support (CAS; see Figure 1).

Caihong, and Xiang Long platforms; and the dress and operational behavior of Chinese aircraft carrier landing deck crews. See Brendan McGarry, "China's Fighters, Drone Look Like U.S. Aircraft," *DefenseTech*, June 20, 2013; Richard Whittle, "New China Drone: Looks Like a Reaper, But . . . ," *Breaking Defense*, September 2, 2015; David Axe, "Go Ahead China—Copy Our Crappiest Warplane," *Daily Beast*, June 5, 2015; Sebastien Roblin, "China Stole This Fighter from Russia—and It's Coming to the South China Sea," *National Interest*, July 24, 2016; Bill Gertz, "Top Gun Takeover: Stolen F-35 Secrets Showing Up in China's Stealth Fighter," *Washington Free Beacon*, March 13, 2014; Andrew Erickson and Gabe Collins, "Chinese Aircraft Carrier Style! Assessing the First Takeoff and Landing," *Wall Street Journal*, November 27, 2012; Marcus Weisgerber, "China's Copycat Jet Raises Questions About F-35," *DefenseOne*, September 23, 2015.

[11] The PLA clearly plans for a large number of possible contingencies and has developed or is developing capabilities, concepts of operation, plans, and supporting structures to meet these, many of which include a prominent role for military aerospace power. Such contingencies include a clash with the United States over Taiwan; a dispute in the East China Sea with Japan and the United States; a war on the Korean Peninsula; a conflict in the South China Sea; and a border war with neighbors like Vietnam or India. This study focused primarily on the role of competition with the United States and the pathways of copying or innovating. It concluded that the main driver for Chinese military aerospace power development is the PLA's view that it needs to be prepared to deter and, if necessary, defeat the United States in a high-end clash. This is not to say that the PLA does not also plan for the military aerospace capabilities and employment concepts to leverage the aerospace domain in other contingencies, some of which may not involve the United States directly or at all. The study did seek to explore the extent to which these other drivers (both in terms of specific contingencies and other actors, including Russia, Japan, India, Taiwan, and European North Atlantic Treaty Organization [NATO] countries) shape PLA aerospace modernization but found little evidence that these serve as key factors shaping China's military aerospace development. The author thanks Michael Lostumbo for encouraging him to make this point more explicit.

Figure 1. Assessing the Degree Of PLA Imitation of U.S. Military Aerospace

Capability/Focus	Chinese Approach Appear to Copy U.S. Approach?	Degree of Similarity?
Air superiority fighters	Yes; J-20, J-31 appear to copy substantial features of F-22, F-35	Greater
AWACS & ISR	Yes; PLA incorporating new airborne C2 and UAV-based ISR capabilities that appear to be modeled on U.S.	
Aerial refueling	Yes; new PLA airframes all include refueling and appear to be based on insights from U.S. air operations	
Training and exercises	Somewhat; PLA early in process of training for jointness and in less scripted scenarios (at night, over water, at distance) modeled on U.S.	Lesser
Bombers and precision-strike	Little; PLA achieves this mission primarily through conventionally-tipped ballistic and cruise missiles, not manned bombers the way the U.S. does, though recent use of bombers for signaling appears somewhat similar to USAF's deployment of strategic assets for signaling	
Close Air Support	No; PLA does not focus much on CAS	

NOTE: Overall assessments are necessarily somewhat subjective and nontechnical, as they are based on open-source materials. AWACS = airborne warning and control system.

It is important to recognize that many of the PLA's efforts in the military aerospace sector focus on fielding of specific capabilities in sufficient quantities to deter the United States from entering a conflict; the PLA would vastly prefer deterrence over actual combat operations. In this sense, the capabilities competition can be regarded as aimed at defeating the United States without actually fighting.

Copying and adapting or, alternatively, innovating capabilities are both valid pathways to this goal, although the lower cost and higher speed of the copying and adapting approach appears to have made it a preferred approach whenever available. However, the PLA copies only aspects of U.S. or foreign practices and capabilities that are relevant to China's goals, then adapts these to local realities. In addition, different services prefer different approaches: The PLAAF tends to copy, while China's missile and space programs are more frequent innovators.

Organization of the Report

Chapter 2 contains a short description of the terminology, methodological approach, and data sources we employed to answer the questions posed earlier. Chapter 3 is a brief overview of the literature on China's military aerospace capabilities, focusing most closely on the period from the mid-1990s to the present. Chapter 4 focuses on Chinese military leadership and the PLA's overall aims. Chapters 5 and 6 examine why and how China competes, copies, and innovates in military aerospace. Chapter 7 concludes with some thoughts on the implications of this research for the broader study of Chinese foreign and security policies, USAF interests and U.S. strategy more broadly, and the understanding of the PLA as an organization.

2. Terminology, Methodology, and Data Sources

As a starting point, it is important to define what it means to ask about whether China's military aerospace seeks to *compete* with the United States, and, if so, in pursuing that goal whether it *copies* or *innovates*.

This study employs China's conceptual approach to military aerospace, which is defined by a focus on "integrated air and space" (空天一体) capabilities and "offensive and defensive operations" (攻防兼备).[1] China's military aerospace capabilities include manned and unmanned fixed- and rotary-wing aviation, ballistic and cruise missiles, and satellites and space-based assets—capabilities that the PLA operates across five services. According to the *Chinese Air Force Encyclopedia*, aerospace capabilities include directed-energy weapons in addition to kinetic weapons.[2]

Military aerospace operations can be divided up by scale (strategic, campaign, and tactical); in terms of area of domain (space-to-space, air-to-space, space-to-ground, air-to-ground, and air-to-air operations); operational method (missile attack/defense, aerospace attack/defense, and air raid/counterraid operations); or missions (integrated air-space information operations, integrated air-space offensives, integrated air-space defensive operations, and integrated air-space support operations).[3]

In terms of the PLA services and branches involved in military aerospace operations, the PLAAF has missions that include lift; ISR; air defense and air attack; and C2. The PLARF is largely responsible for China's nuclear and conventional ballistic and cruise missile capabilities.[4] The PLASSF seems to have inherited the space and counterspace functions of the former General Armaments Department (GAD), which gives it an important role in ISR and strike missions as well as space and counterspace missions. The PLAN operates ship-to-ship missiles and air defenses and has a naval aviation branch (the PLA Navy Air Force [PLANAF]) that is responsible for land-based maritime strike missions and is increasingly fielding carrier-based

---

[1] See, for example, Xi Jinping's speech to the PLAAF on April 14, 2014: "Xi Jinping: Speed up the Construction of a Powerful People's Air Force with Integrated Air-Space and Offensive-Defensive Capabilities" [习近平：加快建设一支空天一体，攻防兼备的强大人民空军], *Xinhua* [新华], April 14, 2014.

[2] Dai Jinyu, Dong Changjun [戴金宇，董长军], "Air-Space Warfare Theory" [空天一体作战理论] in China Air Force Encyclopedia Editorial Advisory Board [中国空军百科全书编审委员会], eds., *Chinese Air Force Encyclopedia*, Vol. 1 [中国空军百科全书，上卷], Beijing: Aviation Industry Publishing House [北京：航空工业出版社], November 2005, pp. 46–47.

[3] "Integrated Air-Space Operations (Part 1)" [空天一体作战（上）], *Xinhua* [新华], November 23, 2015.

[4] There is one exception: China's Type 094 ballistic missile submarines and JL-2 submarine-launched ballistic missiles appear to be under PLAN control.

airplanes as well as helicopters for maritime domain awareness and antisubmarine warfare. Finally, the PLAA operates some helicopters and unmanned aerial vehicles.

In looking at how China approaches the development of military aerospace capabilities across these various domains and services, we define *competing* as a deliberate effort by the Chinese state and military to develop capabilities designed to permit the PLA to keep up with, match, and, if possible, exceed a real or potential rival (in most cases, the United States). In practical terms, this often equates with not being deterred by, deterring, and (if necessary) defeating an adversary (in this case, the United States). In comparison with the other two terms examined next, competing is best seen as a decision that sets a goal for military aerospace power development, with (1) *copying* and *adapting* and (2) *innovating* as two discrete pathways by which to attempt to reach the goal.

Competing is a term that carries two potential inaccurate meanings. The term could convey the impression that China's goal might be simply to match the United States or some other nation in defense aerospace, perhaps for symbolic purposes. As this report argues, this is an incorrect understanding; China's military aerospace power is intended to accomplish a number of goals, but simply matching U.S. capabilities is not one of them. This assessment is confirmed by both a review of primary and secondary source literature on China's military missions and goals and discussions with U.S. subject-matter experts.

A second inaccurate implication is that competition (with the United States or other nations) is the sole driver of PLA aerospace modernization efforts. Fundamental goals and drivers are behind the PLA's aerospace capabilities development, including the defense of the CCP's ruling status and the CCP's definition of China's national interests; interservice rivalries, budgetary battles, and service-specific ambitions; and defense industrial development, which is driven by the interests of science and technology communities linked to specific bureaucratic sponsors. However, the U.S. military's capabilities and operations still shape overall PLA strategy, concepts of force employment, capabilities generation, and other aspects of military operations. As noted in a recent study, "[t]he PLA's doctrinal and training revisions have been made purely in response to observations of the performance of foreign militaries, particularly the U.S. military."[5]

Copying, as the term is used here, refers to China's introduction of foreign technology designs into the PLA following their acquisition through the theft or reverse engineering of foreign military-industrial intellectual property.[6] Copying can also refer to the incorporation of

---

[5] Roger Cliff, "PLA Threat Assessments, Equipment Modernization, Doctrine and Training: 1993–2013," conference paper for CAPS-RAND-NDU International Conference on PLA Affairs, 2013.

[6] On China's technology theft, see Phillip C. Saunders and Joshua K. Wiseman, *Buy, Build, or Steal: China's Quest for Advanced Military Aviation Technologies—Institute for National Strategic Studies China Strategic Perspectives*, No. 4, Washington, D.C.: National Defense University Press, 2011; and William C. Hannas, James Mulvenon, and Anna B. Puglisi, *Chinese Industrial Espionage: Technology Acquisition and Military Modernization*, New York: Routledge Press, 2013.

operating patterns and lessons after a period of observation of foreign militaries' behavior and activities. When China copies something that the United States, Russia, or another country does, it is of course not an exact imitation; Chinese practice is generally to adopt and then adapt foreign technologies and practices to the needs and capabilities of Chinese military aerospace. In practical terms, this means that substantial elements of Chinese capabilities or behavior may be identifiably modeled on USAF or other foreign air forces' hardware or operating practices while possessing other features that are distinctively Chinese in origin. Prominent examples of copying include some of the external design features of U.S. advanced fighters and remotely piloted aircraft.

In contrast, innovation refers to situations in which the PLA or the Chinese defense industry has developed a new technology substantially through its own efforts or has solved a military problem or task in a novel way through indigenous operational or organizational responses. Innovation can occur in relation to hardware, organization, operations, or any combination thereof. While Tai Ming Cheung of the University of California at San Diego has offered the helpful definition that innovation can be thought of as separated into two types—"defense innovation" and "military innovation"—this study adopts a definition of innovation as "the transformation of ideas and knowledge into new or improved products, processes and services for military and dual-use applications . . . [as well as other steps] intended to enhance the military's ability to prepare for, fight, and win wars."[7] Because most military innovation appears to involve incremental changes over past hardware, organizational, or operational practices, this definition is useful insofar as it does not define innovation exclusively in terms of major breakthroughs; this would set the bar too high. Examples of innovation include the PLA's use of missiles in place of manned bombers and the Chinese approach to defense industrial prototyping and transformation of conceptual designs into platforms that have achieved initial operational capability.

One final definitional point is in order. While the terminological descriptions offered here are intended to help define thinking about PLA aerospace power, they are also heuristic devices for understanding a complex reality, rather than clear-cut categories. In practice, any given piece of PLA aerospace hardware, any operational practice, any doctrinal approach or organizational arrangement can embody both copied elements and innovation. Furthermore, innovation can occur in the employment of copied hardware, and an innovative application of military technology to solve a problem can involve the use of hardware stolen or copied from a foreign power. The very notions of copying and innovating are probably best thought of as a spectrum rather than a binary.

In terms of our approach to investigating China's military aerospace power development, we initially turned to written Chinese sources and other materials that were based on official U.S.

---

[7] Tai Ming Cheung, ed., *Forging China's Military Might: A New Framework for Assessing Innovation*, Baltimore, Md.: Johns Hopkins University, 2013, p. 3.

observations of Chinese military operations. However, we found that once we delved below the top-level mission sets and goals for aerospace development, Chinese sources did not tend to address the more-granular policy questions at the heart of this study. As such, it was ultimately necessary to expand our data inputs by drawing on a wide variety of secondary sources by Western academics, think-tank experts, and U.S. government reports on the PLA, especially those focused on organization, decisionmaking, and the development of military power.[8]

I therefore conducted more than a dozen open-ended, face-to-face interviews with leading American subject-matter experts specializing in various aspects of the Chinese armed forces—particularly those focused on air power, ballistic and cruise missiles, and space capabilities—to fill in gaps in our understanding of the priorities and ways of thinking about military aerospace that China holds. I asked our interviewees to comment on whether they see China as competing with the United States; sought their input on whether and, if so, when China copies the United States in military aerospace; inquired about areas where China innovates; and generally sought any other information that might be relevant, including historical and comparative points, organizational insights, and descriptions of operational activities. The interviews provided insight into the goals and factors that shape the decisionmaking process whereby China sets its priorities in the development of military aerospace power. Interviewees came from a range of prominent U.S. think tanks, academic institutions, and government organizations and generally had two or more decades of experience focusing on the PLA. All of the interviews were conducted on a promise of anonymity and were done at the unclassified level; additionally, all of the information from the interviews and all other sources referenced in this report are unclassified.

---

[8] See U.S. Department of Defense, *Annual Report to Congress—Military and Security Developments Involving the People's Republic of China*, Washington, D.C., 2013–2017; Dennis Blasko, "Recent Developments in the Chinese Army's Helicopter Force," *Jamestown Foundation China Brief*, Vol. 17, No. 8, June 9, 2017; Roger Cliff, *China's Military Power: Assessing Current and Future Capabilities*, Cambridge, UK: Cambridge University Press, 2015; Eric Heginbotham, Michael Nixon, Forrest E. Morgan, Jacob L. Heim, Jeff Hagen, Sheng Li, Jeffrey Engstrom, Martin C. Libicki, Paul DeLuca, David A. Shlapak, David R. Frelinger, Burgess Laird, Kyle Brady, and Lyle J. Morris, *The U.S.–China Military Scorecard: Forces, Geography, and the Evolving Balance of Power, 1996–2017*, Santa Monica, Calif.: RAND Corporation, RR-392-AF, 2015; Richard P. Hallion, Roger Cliff, and Phillip C. Saunders, *The Chinese Air Force: Evolving Concepts, Roles, and Capabilities*, Washington, D.C.: National Defense University Press, 2013; Michael S. Chase, Jeffrey Engstrom, Tai Ming Cheung, Kristen A. Gunness, Scott Warren Harold, Susan Puska, and Samuel K. Berkowitz, *China's Incomplete Military Transformation: Understanding the Weaknesses of the PLA*, Santa Monica, Calif.: RAND Corporation, RR-893-USCC, 2015; Michael S. Chase, Kristen A. Gunness, Lyle J. Morris, Samuel K. Berkowitz, and Benjamin S. Purser, III, *Emerging Trends in China's Development of Unmanned Systems*, Santa Monica, Calif.: RAND Corporation, RR-990-OSD, 2015; Kevin Pollpeter, Eric Anderson, Jordan Wilson, and Fan Yang, *China Dream, Space Dream: China's Progress in Space Technologies and Implications for the United States*, Washington, D.C.: U.S.–China Economic and Security Review Commission, 2015; and Tai Ming Cheung, *Fortifying China: The Struggle to Build a Modern Defense Economy*, Ithaca, N.Y.: Cornell University Press, 2008.

3. Rapid Recent Development from a Low Post–Cold War Baseline

China's rapid development of advanced military capabilities has repeatedly surprised U.S. analysts and policymakers, a viewpoint expressed by both former Secretary of Defense Robert Gates and Admiral Robert Willard, the former commander of U.S. Pacific Command.[1] From the mid-1990s onward, experts on the PLAAF, the PLARF (formerly the PLA Second Artillery Force, or PLASAF), and Chinese space programs have assessed China's prospective military aerospace capabilities; over time, the PLA's actual military aerospace development has repeatedly met or surpassed the most optimistic assessed prospects.

For example, a 1995 study of PLAAF development by the RAND Corporation noted "a large number of deficiencies, [including in the realms of] . . . surveillance capabilities, and offensive power" and argued that "China's aviation manufacturing sector is in a poor position to help provide the PLAAF with much-needed advanced weapon systems." As such, it concluded, "the PLAAF will [not] emerge as an appreciably more formidable player in the global balance of airpower" by 2005, but could do so by 2015, "provided [it] continues to reform, develops its aerospace industrial infrastructure, and enjoys greater access to resources."[2] A 2005 RAND study found that, as a result of increases in funding, enhanced reliance on market incentives, expanded access to foreign defense hardware (primarily from Russia and Israel), and a series of structural reforms to the defense procurement system, "limited signs of increasing progress" could be seen in the Chinese military aviation sector, leading the authors to conclude that

> China's defense industry now has the potential to become more competitive with the defense industries of the world's advanced military powers in key sectors within a moderate (10–20 years) amount of time.[3]

A separate RAND study of the cross-Strait balance, published in 2000, found that the PLA's improvements to its air and missile portfolio were giving China a much greater capacity to credibly threaten Taiwan.[4]

Nine years later, a follow-up examination of the military balance across the Taiwan Strait concluded that China's improved air and missile assets had shifted the cross-Strait balance so far

---

[1] James Holmes and Toshi Yoshihara, "Underestimating China," *The Diplomat*, January 17, 2011.

[2] Kenneth W. Allen, Glenn Krumel, and Jonathan D. Pollack, *China's Air Force Enters the 21st Century*, Santa Monica, Calif.: RAND Corporation, MR-580-AF, 1995, pp. xx–xxi.

[3] Evan S. Medeiros, Roger Cliff, Keith Crane, and James Mulvenon, *A New Direction for China's Defense Industry*, Santa Monica, Calif.: RAND Corporation, MG-334-AF, 2005, pp. xvii–xviii, xxiii–xxiv.

[4] David A. Shlapak, David T. Orletsky, and Barry A. Wilson, *Dire Strait? Military Aspects of the China-Taiwan Confrontation and Options for U.S. Policy*, Santa Monica, Calif.: RAND Corporation, MR-1217-SRF, 2000.

that the PLAAF had a reasonably good prospect of achieving "air superiority over Taiwan and the strait" as well as threatening nearby U.S. military facilities and assets, meaning that "the air war for Taiwan could essentially be over before much of the Blue [i.e., U.S. and Taiwan] air forces have even fired a shot."[5] Reflecting the growing recognition inside China of the importance of airpower, a 2011 RAND study found that China's military appeared to have concluded that

> the struggle for dominance of the battlefield will increasingly consist of an integrated struggle for air, space, information, and electromagnetic (and even computer network) superiority. Acquiring air superiority is considered a prerequisite in a variety of operations involving all services.[6]

Analysts have noted that China's efforts to build up and modernize its military capabilities appear to have accelerated substantially in the wake of the 1991 Gulf War, especially after the 1995–1996 Taiwan Strait crisis and the 1997 announcement of the U.S.–Japan Revised Defense Guidelines. The fear that the PLA would be incapable of posing a serious deterrent threat to Taiwan's permanent political separation from China spurred an innovative approach to deterring this outcome based on expansion of the PLA's portfolio of short-range ballistic missiles and extensive improvements in space-based reconnaissance and targeting capabilities designed to make conventional missiles much more accurate and operationally valuable.

Perhaps more consequential than any of these factors, however, was the accidental bombing of the Chinese embassy in Belgrade in 1999. The experience of being unable to respond to the U.S. military was deeply humiliating for the PLA and CCP. One expert interviewed for this study noted that the PLA's "vision of creating 'assassin's mace' [or trump card] weapons that can 'look far, shoot far, and shoot accurately' was driven by the 1999 embassy bombing."[7]

The U.S. military's domination of Iraq in the 1991 Gulf War and the use of advanced helicopter and air power in the 1999 Kosovo conflict not only demonstrated the value of advanced airpower but also highlighted the gaps between the PLA and the U.S. military, spurring China's initial shift towards a greater emphasis on air power, informatization, and network-centric (or in Chinese terms, "informationalized" [信息化]) warfare. The PLA's attempts to procure airborne early warning and control platforms from Israel in the late 1990s

> demonstrated a recognition that air power is not just about fighters but instead about [the linking together of a variety of] capabilities in a reconnaissance-strike kill-chain complex. This represented a maturation of the PLA's thinking about air

---

[5] David A. Shlapak, David T. Orletsky, Toy I. Reid, Murray Scot Tanner, and Barry Wilson, *A Question of Balance: Political Context and Military Aspects of the China-Taiwan Dispute*, Santa Monica, Calif.: RAND Corporation, MG-888-SRF, 2009.

[6] Roger Cliff, John F. Fei, Jeff Hagen, Elizabeth Hague, Eric Heginbotham, and John Stillion, *Shaking the Heavens and Splitting the Earth: Chinese Air Power Employment Concepts in the 21st Century*, Santa Monica, Calif.: RAND Corporation, MG-915-AF, 2011, p. xvii.

[7] RAND interview 11, July 2017.

> power, the aerospace domains, and jointness in sensors and strike capabilities. [And with the advent of a focus on integrated air and space joint operations] the Chinese have recognized that airpower is a necessary, but not sufficient, capability for war-winning.[8]

The 2001 toppling of the Taliban regime in Afghanistan and the initial successes of the 2003 overthrow of the Saddam Hussein regime in Iraq further demonstrated the influential role of air power in modern warfare. The ability to dominate the skies over both countries allowed the United States to strike enemy key point targets with impunity, ensure situational awareness and assist with force protection, deliver forces and supplies virtually throughout the two countries, and seize and maintain information dominance throughout the two battlespaces. In the case of Afghanistan, the United States was even able to topple a government without introducing substantial ground combat power by combining a campaign to suppress enemy air defenses with a strategic bombing campaign and a series of targeted insertions of Central Intelligence Agency and Special Forces capabilities that linked up with indigenous opposition groups.[9] These campaigns made a deep impression on Chinese military observers and appear to have reshaped Chinese thinking on the importance of military aerospace.[10] Perhaps reflecting this, in 2004, the PLAAF articulated its first-ever service-specific strategic concept and saw its service chief appointed to the CMC for the first time in CMC history; with these changes, the importance of air power came to be recognized and respected more within the PLA.[11]

Alongside these changes, during the mid-to-late 1990s, China initiated programs designed to produce advanced fighters, such as the J-20 and J-31 fighters; in the early-to-mid 2000s, China began research and design on its early-generation unmanned aerial vehicles. From 2012, when Xi Jinping took the helm of the CMC, the PLAAF has seen its influence grow further and its prominence increase with the roll-out of new advanced fighters, such as the J-20 and J-31. In addition, former PLAAF Commander Xu Qiliang was promoted to the post of CMC vice-chairman, while General Ma Xiaotian, his successor as PLAAF commander, joined the CMC as a full member, giving the PLAAF two full members to advocate on behalf of the service's interests. Reorganization and personnel changes on the CMC that were enacted at the 19th Party Congress in October 2017, however, left the PLAAF Commander's post in the hands of a three-

---

[8] RAND interview 12, August 2017.

[9] See Gary Schroen, *First In: An Insider's Account of How the CIA Spearheaded the War on Terror in Afghanistan*, New York, N.Y.: Presidio Press, 2005; Gary Berntsen with Ralph Pezzullo, *Jawbreaker: The Attack on Bin Laden and al Qaeda—A Personal Account by the CIA's Key Field Commander*, New York, N.Y.: Crown Publishing, 2005.

[10] These campaigns also further underlined the importance of informationalized capabilities to Chinese military observers. See Shou Xiaosong [寿晓松], ed., *Science of Strategy* [战略学], Beijing: Military Science Publishing House [军事科学出版社], 2013, p. 97.

[11] See Chase and Garafola, 2015, p. 6; on the promotion of the PLAAF commander to the CMC, see Cheng Li and Scott W. Harold, "China's New Military Elite," *China Security*, Vol. 3, No. 4, Fall 2007, pp. 62–89.

star general officer and reduced the size of the CMC, leaving only one PLAAF general on the committee (Xu Qiliang, who continues to serve as vice-chairman at the time of writing).

It is important to bear in mind, however, that while China's decisions to build up its military aerospace capabilities from the mid- to late-1990s onwards were substantially influenced by changes in the country's external environment, they are not the only factors worth pointing out; domestic developments also played a critical role. One particularly important factor was the rapid growth of China's economy after 1992; this provided substantial resources for a growing defense budget. Other key developments were the 1998 reform of the Committee on Science and Technology in National Defense (COSTIND) and the establishment within the CMC of the GAD.[12] These changes to how China funded and organized its defense industrial base and the prioritization of the needs of the PLA over the interests of defense industrial firms in procurement substantially improved China's ability to generate military power.

The 1998 changes to COSTIND and the establishment of the GAD served to link PLA assessments of the external threat environment more closely to the development goals of the Chinese defense industrial base and procurement. As one interviewee noted,

> the PLA was basically shut out of the acquisition process until the establishment of the GAD . . . GAD was the solution to bring PLA warfighters into the procurement process. Acquisition and procurement was no longer a jobs program for defense industrial firms; [in the years following the 1998 reforms, China moved] from a "technology push" model to [an approach based on] geographical and technological threat assessments.[13]

Research in the years that followed substantiated the importance of these changes for improving the Chinese defense industry's ability to deliver technology solutions in which the PLA was interested.[14]

Despite these post-1998 improvements, the military that China fields today continues to be shaped by historical factors that deeply affect its composition, doctrine, and structure. This is particularly true in the realm of military aerospace. As one interviewee for this study noted,

> China's history with air power is totally different than that of the [United States], Germany, Russia and other great powers. Who was the first Chinese pilot? When did the Chinese first assemble an airplane themselves—the 1930s, perhaps? History matters, and a country's air power roots explain a lot [about its identity and vision of its role.] The [PLAAF] has no lineage, history, or traditions to draw upon. Aerospace development can generally be broken into three constitutive elements: airplanes, missiles and space capabilities; civilian airliners and air transport companies; and airfields, air traffic controllers and aerospace firms.

---

[12] An Ti [安替], "CMC's New Weapons and Armaments Administration Regulations Advance Modernization of Chinese Military Equipment" [军委武器装备管理新条例推动中国军备现代化], *People's Daily Online* [人民网], January 21, 2003.

[13] RAND interview 10, July 2017.

[14] Medeiros et al., 2005; Cheung, 2008; James Mulvenon and Rebecca Samm Tyroler-Cooper, *China's Defense Industry on the Path of Reform*, Washington, D.C.: U.S.–China Economic and Security Review Commission, 2009.

> China had almost none of these until the 1950s [i.e., much later than other advanced nations].[15]

This view of the importance of history was echoed by another interviewee, who commented that

> [t]he [PLAAF] doesn't have traditions to look back on and so has to adapt foreign military theory to China's realities. This often means that the PLA lacks confidence [in its own capabilities].[16]

A third expert commented that because the PLA

> lacks experience of their own, they have to copy a lot. In terms of air operations, they only learn from the United States. [The United States] is also the only opponent they build against. They don't copy [to match the United States] but to be able to beat us.[17]

While the PLAAF is evolving from a low baseline, it nonetheless has developed substantially in recent years beyond its traditional limited role of air defense over the Chinese homeland and has overcome the associations that put the entire service under a cloud following an alleged coup d'etat against Mao Zedong in 1971.[18] However, the PLAAF still provides a stark contrast to China's strategic missile and space programs, which were developed indigenously and protected from political and economic infighting, even during the Cultural Revolution. These programs, together with China's nuclear weapons development project, were given resources, support, and time to develop; as a result, they are some of the most innovative parts of the PLA.[19] Meanwhile, the PLAAF was populated with Soviet hardware and saddled with the legacies of the Soviet Union and the failed coup d'etat. In the 1990s, the PLAAF again turned to Russia for hardware, as the Chinese defense industrial base could not produce advanced fighters. Even today, China remains incapable of producing military-grade fighter engines.[20]

---

[15] RAND interview 12, August 2017.

[16] RAND interview 2, July 2017.

[17] RAND interview 5, July 2017.

[18] At the time of Marshal Lin Biao's death under mysterious circumstances in 1971, he was China's minister of national defense and Mao's designated successor. Lin allegedly concluded that Mao had turned against him and attempted to assassinate Mao in a coup d'etat associated with the PLAAF. The plot was uncovered, and Lin attempted to flee the country in a hurry using a plane that was only partially fueled; the plane ultimately crashed in Outer Mongolia.

[19] The missile, satellite, and atom and hydrogen bomb programs were grouped together as strategic capabilities protected from political interference. This was known as the "two bombs, one satellite" (两弹一星) initiative. See John Wilson Lewis and Xue Litai, *China Builds the Bomb*, Stanford, Calif.: Stanford University Press, 1988; and Evan A. Feigenbaum, *China's Techno-Warriors: National Security and Strategic Competition from the Nuclear to the Information Age*, Stanford, Calif.: Stanford University Press, 2003.

[20] Industrial espionage was also a part of China's solution to military aerospace development, escalating in importance from the early 1990s onwards.

Currently, Chinese military aerospace power is a mix of legacy capabilities derived from the Cold War era; hardware procured from Russia and Ukraine; copies and knock-offs of Russian airframes produced by kit or through reverse engineering; a small but growing number of fourth- and fifth-generation fighter-bombers that appear to have been developed based in part on stolen designs for U.S. airframes; advanced, but nonstealthy, ballistic and cruise missiles; and a growing, largely indigenously developed portfolio of space and antispace capabilities.

Before the 1998 establishment of the GAD, the PLA was constantly battling with various defense industrial interests over regular weapons programs. Even after the establishment of the GAD, Presidents Jiang Zemin and Hu Jintao found it necessary to set up various special weapons initiatives to insulate high-priority research and development programs from bureaucratic and commercial capture. As one expert explains, "this accounts for the black and white nature of the Chinese military, which has some good and many very poor quality weapons programs."[21] In an effort to fix these problems, China continued to pursue reforms of its defense industrial base, procurement processes, and overall military structure throughout the 2000s, although these efforts were often piecemeal, and major reforms were frustrated until the Xi Jinping era began in 2012. Under Xi Jinping, a number of important steps have been taken that further prioritize the development of military aerospace power; these are detailed in the chapter on innovation.

In short, from a weak base, China's military aerospace has grown substantially more capable in the past two decades, fueled in large part by a perception that the country needs to develop modern military aerospace capabilities to meet the military missions that the CCP has identified for the PLA. Given the very different histories, political contexts in which they exist, capabilities, and national interests they are intended to support, it is perhaps not surprising that the PLAAF and the other elements of China's military aerospace exhibit some differences from their U.S. counterparts; they also appear, however, to have substantial similarities, not all of which are derived purely from requirements imposed by the physical challenges of advanced aerospace operations.

---

[21] RAND interview 10, July 2017.

4. PLA Aims to Defeat, Not Merely Compete with, U.S. Military

When the Chinese military leadership decides on the development of military aerospace power, a number of factors shape their decision calculus. Important factors include the PLA's priority mission sets,[1] the resources available for aerospace development, and the capabilities of the Chinese defense industrial sector (with assistance from foreign defense technology, procured through aboveboard means or covertly). The PLA's need to keep up with, match, and (if possible) exceed U.S. capabilities reflects its assessment of the key operational challenges to the accomplishment of important missions, such as the defense of Chinese airspace, the prosecution of a conflict over Taiwan, or the projection of power in the East and South China Seas inside the first island chain to prosecute claims to disputed land features and maritime spaces.[2] As noted previously, China competes to deter the United States and other adversaries from confronting China or preventing it from achieving its policy goals; if necessary, the PLA should be able to defeat these adversaries if they do choose to contest China's policy moves. The PLA appears not to compete in certain areas because it does not need certain capabilities to accomplish its directed mission, or it has other means to address the military problem at hand. This chapter describes the role of foreign (and especially U.S.) military power as a driver of PLA capabilities acquisition and military aerospace power development.

The PLA has a long history of engaging in military diplomacy to absorb technology and operational lessons through contact with foreign militaries;[3] the PLA studies the U.S. and other militaries' experiences carefully.[4] PLA analysts may draw the wrong lessons, or their lessons may not be effectively incorporated within the PLA, but the PLA still is making a deliberate and

---

[1] China's military mission sets are described in Chase et al., 2015, especially pp. 25–42.

[2] In the course of the research for this study, we also explored the extent to which competition with other advanced powers or rivals—including Russia, Japan, India, NATO, Israel, and Taiwan—shapes PLA military aerospace power development. The key takeaway from our reviews of written sources and interviews with subject-matter experts was that all other militaries pale in comparison to the U.S. military as the pacing factor spurring Chinese military aerospace development. While the PLA clearly learns from (and in some cases imports or steals from) other advanced militaries, the main rival it plans against is the U.S. military.

[3] On China's view of military diplomacy, see Kenneth W. Allen and Eric A. McVadon, *China's Foreign Military Relations*, Washington, D.C.: Henry L. Stimson Center, 1999; Kristen Gunness, "China's Military Diplomacy in an Era of Change," conference paper prepared for National Defense University's "China's Global Activism: Implications for US Security Interests," June 20, 2006; Matsuda Yasuhiro, "An Essay on China's Military Diplomacy: Examination of Intentions in Foreign Strategy," *National Institute of Defense Studies Security Reports*, No. 7, December 2006, pp. 1–40; Shannon Tiezzi, "3 Goals of China's Military Diplomacy," *The Diplomat*, January 30, 2015.

[4] Andrew Scobell, David Lai, and Roy Kamphausen, eds., *Chinese Lessons from Other People's Wars*, Carlisle, Pa.: U.S. Army War College Press, 2011, pp. 4–8.

persistent effort to improve its understanding of major trends in warfare, especially those related to the capabilities and practices of the United States.

Commenting upon Chinese learning from foreign experience, one subject-matter expert noted that "the Chinese definitely observe the organizational structure, technology, and behavior of all foreign militaries," a remark that a separate expert expanded upon by explaining that

> the PLA studies us not to [compete with us or] copy us, but to beat us. They look at how we go to war and they plan against that, structuring themselves to beat the U.S. military and the [USAF] specifically.[5]

Other experts have similarly noted that the PLA appears to be studying the changing capabilities, operational practices, staffing, training, doctrine, basing, and force posture of the U.S. military more closely than all others. While Chinese defense leaders see the United States as powerful, they have not adopted a more compliant posture but rather seek to address military shortcomings so as to stand up the United States.[6]

Recent Chinese writings consistently reflect the view that the primary threat to the accomplishment of the CCP's "China Dream" under Xi Jinping, as well as a major driver shaping the PLA's development of capabilities, is the presence, posture, and capabilities of the armed forces of the United States and the U.S. alliance network in the Asia-Pacific.[7] Lyle Goldstein of the U.S. Naval War College, reviewing Chinese writings on the U.S. rebalance to the Asia-Pacific under the Obama Administration, characterizes Chinese authors as seeing the rebalance as "worse than containment" and finds that "rivalry discourse has become conventional wisdom and even modish in China."[8] Indeed, the 2013 edition of the PLA's Science of Military Strategy describes the basic goal of the military strategy of the United States as being focused on "expansion and hegemony."[9] Adam Liff of Indiana University sees this view as having deep roots in the history of the People's Republic of China, but also as taking on increasing prominence under Xi Jinping, with China expressing a growing frustration with the continued existence of the U.S. alliance network, a set of relationships Beijing characterizes as a "Cold War relic" and an obstacle to the achievement of China's regional ambitions and security

---

[5] RAND interview 2, July 2017; RAND interview 7, July 2017.

[6] On China's adaptation to shifts in U.S. defense posture and policy, see James C. Mulvenon, Murray Scot Tanner, Michael S. Chase, David Frelinger, David C. Gompert, Martin C. Libicki, and Kevin L. Pollpeter, *Chinese Responses to U.S. Military Transformation and Implications for the Department of Defense*, Santa Monica, Calif.: RAND Corporation, MG-340-OSD, 2006.

[7] Timothy R. Heath, Kristen Gunness, and Cortez A. Cooper, III, *The PLA and China's Rejuvenation: National Security and Military Strategies, Deterrence Concepts, and Combat Capabilities*, Santa Monica, Calif.: RAND Corporation, RR-1402-OSD, 2016.

[8] Lyle J. Goldstein, "How China Views America's Moves in the Asia-Pacific: Worse than Containment," *National Interest*, October 29, 2014.

[9] Shou, 2013, p. 52.

goals.[10] Jae Ho Chung of Seoul National University has detailed a similar finding with respect to China's view of the U.S. alliance with the Republic of Korea specifically.[11]

A leading proponent of this perspective that the United States is the main stumbling block to China's national rejuvenation has been PLA author Liu Mingfu. In his 2010 book *The China Dream* (a volume whose themes subsequently appeared frequently in the speeches of Chinese President Xi Jinping), Liu argued that "[i]t has been China's dream for a century to become the world's leading nation," a vision that Liu argues requires the achievement of a corresponding "strong army dream" and the overcoming of America's military preeminence. Liu expands on this by arguing that

> [t]he goal of China's military rise is to make the United States unable to afford to contain China. With this military rise, China will be able to prosper without being peacefully contained by the U.S., and will also be able to contain the U.S. *China's military strength has to be more powerful than any rivals in the world* to the degree and level that no nation can contain China's rise.[12]

Similar sentiments were expressed in a 2013 teaching video titled *Silent Contest* (较量无声), which was created by the PLA's General Staff Department, PLA National Defense University, and the Chinese Academy of Military Sciences; the video was intended to warn PLA officers about the need to guard against psychological warfare and ideological contamination by the United States.[13] While previously some analysts might have argued that these voices constitute the hardline end of Chinese military's spectrum of views of the United States and the ambitions the PLA has for the region, China's aggression against numerous neighbors and actions to realize its ambitions to remake the geopolitical environment in Asia in recent years suggest that these views—if they were ever marginal—have moved much more to the center of policy debates under Xi Jinping.[14]

---

[10] Adam P. Liff, "China and the U.S. Alliance System," *China Quarterly*, April 24, 2017, pp. 1–29.

[11] Jae Ho Chung, "China's Evolving View of the Korean-American Alliance, 1953–2012," *Journal of Contemporary China*, Vol. 23, No. 87, 2014, pp. 425–442.

[12] Liu Mingfu,"The World Is Too Important to be Left to America," *The Atlantic*, June 4, 2015. Italics added.

[13] Jane Perlez, "Strident Video by Chinese Military Casts U.S. as Menace," *New York Times*, October 31, 2013.

[14] David Shambaugh described a range of Chinese elite thinkers' opinions on relations with the outside world shortly before Xi Jinping took power. See David Shambaugh, "Coping with a Conflicted China," *Washington Quarterly*, Vol. 34, No. 1, Winter 2011, pp. 7–27. Similarly, Andrew Scobell and Scott W. Harold found a substantial range of opinions: See Andrew Scobell and Scott W. Harold, "An Assertive China? Insights from Interviews," *Asian Security*, Vol. 9, No. 2, 2013, pp. 111–131. Some observers expressed a view that the public "hawks" in the PLA might be largely performing a role focused on external propaganda and image management (without necessarily reflecting an effort to move the policy debate within the CMC or China's broader policymaking community); see David Lague, "Special Report: China's Hawks Take the Offensive," Reuters, January 16, 2013; Andrew Chubb, "Propaganda, Not Policy: Explaining the PLA's 'Hawkish Faction' (Part One)," *Jamestown Foundation China Brief*, Vol. 13, No. 15, July 25, 2013; and Andrew Chubb, "Propaganda as Policy? Explaining the PLA's 'Hawkish Faction' (Part Two)," *Jamestown Foundation China Brief*, Vol. 13, No. 16, August 9, 2013. More-recent scholarship has reflected a reduction in debate and an emerging consensus in China in favor of a more assertive or aggressive approach to the outside world. See Oriana Skylar Mastro, "Why Chinese Assertiveness is

Among China's security policy priorities and ambitions, apart from the defense of the ruling status of the CCP, the most important goal is to achieve the absorption of Taiwan—by force if necessary. This goal requires that the PLA be capable of overcoming the resistance posed by the armed forces of the Republic of China (Taiwan) as well as any intervention by outside forces, presumably the United States (although other nations, such as Japan, could intervene as well).[15] Experts on the Chinese military surveyed for this study agreed that

> the primary driver of [the PLA's military power development] is the need to defeat the [United States] in a Taiwan scenario. No other country threat really drives PLA aerospace development [to the same extent]. And in designing their capacity to defeat [U.S.] intervention in a Taiwan scenario, they necessarily have to compete with the [United States].[16]

As another expert noted,

> any capability the [United States] is getting, the PLA wants to get. This is not the case with any other military; they will acquire useful capabilities from other countries whenever possible, but there is less compulsion or urgency to this. The [United States] is their most advanced and capable adversary and therefore it must be copied to some extent so as to keep up or if possible overtake it as well as to avoid potential defeat.[17]

Competition with the United States "has driven the key strategic development [initiatives in PLA aerospace power] over the last two decades," and although the Chinese military rarely says so in its public commentary, since the end of the 1990s "they have viewed the [United States] as their main strategic adversary," notes another PLA expert.[18] In keeping with this view, another interviewee noted that

> the PLA's mission sets affect the weight they accord to a particular capability [but to the best of their ability they try] to match everything the [United States] tries to build; it is just a matter of to what degree they are prioritizing it. [Basically] any capability the [United States] is getting, the PLA wants to get. This is not necessarily the case with any other military.[19]

Here to Stay," *Washington Quarterly*, Vol. 37, No. 4, 2014, pp. 151–170; Yan Xuetong, "From 'Keeping a Low Profile' to 'Striving for Achievement,'" *Chinese Journal of International Politics*, Vol. 7, No. 2, Summer 2014, pp. 153–184. Kurt Campbell and Robert Blackwill, *Xi Jinping on the Global Stage: Chinese Foreign Policy Under a Powerful but Exposed Leader—Council Special Report No. 74*, New York, N.Y.: Council on Foreign Relations, February 2016, offers a compelling explanation for why China's foreign policy has turned more hardline by linking it to the policy challenges and personal incentives confronting an ambitious and ideologically motivated Chinese leader looking to consolidate and defend his political authority.

[15] RAND interview 9, July 2017.

[16] RAND interview 7, July 2017.

[17] RAND interview 2, July 2017.

[18] RAND interview 10, July 2017. The reference to "the end of the 1990s" highlights the PLA's sense of vulnerability and ineffectualness following the 1995–1996 Taiwan Strait crisis; the need to respond to the 1997 U.S.–Japan Revised Defense Guidelines; and the humiliating experience of the Belgrade embassy bombing.

[19] RAND interview 2, July 2017.

In regarding a Taiwan Strait contingency as the main threat to plan against and the possible intervention of the U.S. armed forces as the greatest obstacle to the achievement of China's goals, the PLA and the CMC must then decide how best to meet this challenge. There are a limited number of options available to China in its quest to generate military power. It can seek to buy hardware off the shelf from foreign providers, although this option is severely constrained, as China has been under an arms embargo by the United States and its allies since the 1989 Tiananmen Square massacre.

There are only a few other producers of leading edge military hardware that China can seek to buy from. One has been Russia, which has sold China some advanced aerospace capabilities, including fixed-wing airframes (Su-27, Su-30, and Su-35 fighters); Il-76 heavy lift transporters and Il-78 aerial refueling tankers; the SS-N-22 Sunburn antiship cruise missile; and S-300 and S-400 class SAM systems, among other capabilities.[20] A second foreign source of military capabilities for the PLA has been Ukraine, which has provided China with turbofan engines, refueling tankers, An-225 heavy lift transporters, and upgrades for fighter jets.[21] A third has been Israel, which transferred HARPY high-speed antiradiation drones and the plans for the Lavi-class fighter (subsequently reborn as the J-10). Israel also attempted to transfer to the PLA the advanced PHALCON AWACS platform, but the deal was ultimately canceled under U.S. pressure.[22] Finally, some Chinese aircraft field French helicopter technologies and British engines and early warning radar.[23]

After receiving such technologies, sometimes in limited quantities, China has often sought to reverse engineer them to produce them indigenously, a process Chinese analysts today describe as "IDAR," or "introduce, digest, absorb, and re-innovate" (引进, 消化, 吸收, 再创新).[24] When foreign purchases are not an option, China has generally sought to steal foreign technology designs or observe foreign practices with an eye toward copying and adapting them to suit the PLA's needs.[25] Alternatively, when purchase or stealing/copying are not feasible, China has

---

[20] Siemon T. Wezeman, "China, Russia, and the Shifting Landscape of Arms Sales," Stockholm International Peace Research Institute, July 5, 2017.

[21] Will Ponomarenko, "China Sees Ukraine as Alternative to Russia in Arms Trade, Expert Believes," *Kyiv Post*, May 14, 2017.

[22] James Mann, "U.S. Says Israel Gave Combat Jet Plans to China," *Los Angeles Times*, December 28, 1994; Islam Ayyadi and Mohammad Kamal, "China-Israel Arms Trade and Cooperation: History and Policy Implications," *Asian Affairs*, Vol. 47, No. 2, 2016, pp. 260–273.

[23] Stuart McMillan, "Europe's Arms Trade with China," *ASPI Strategist*, February 11, 2014.

[24] Tai Ming Cheung, "The Role of Foreign Technology in China's Defense Research, Development, and Acquisition Process," *Study of Innovation and Technology in China Policy Brief*, January 5, 2014; see also Wang Xiaohong [王晓红], "The Starting Point for Introduce, Digest, Absorb and Re-Innovate" [引进消化吸收再创新的出发点], *People's Daily Online* [人民网], April 3, 2006.

[25] On China's options for procuring advanced military aviation technologies, see Phillip C. Saunders and Joshua K. Wiseman, *Buy, Build, or Steal: China's Quest for Advanced Military Aviation Technologies—Institute for National Strategic Studies China Strategic Perspectives*, No. 4, Washington, D.C.: National Defense University Press, 2011;

sought to innovate solutions to its military problems. These two options—copying and innovation—are explored further in the next two chapters.

---

and William C. Hannas, James Mulvenon, and Anna B. Puglisi, *Chinese Industrial Espionage: Technology Acquisition and Military Modernization*, New York, N.Y.: Routledge Press, 2013.

5. When Does China Copy Foreign Militaries?

The recent PLAAF focus on what one Chinese commentator has dubbed "bright eyes, strong fists, and long arms" (亮眼、重拳、长臂), with its similarity to the USAF's "global vigilance, global reach, global power" formula, is clearly focused on building greater capacity to gain situational awareness, deploy assets, and project power over greater distances.[1] Whether China is copying the U.S. military or whether China's interests are evolving in ways that force it to adopt strategies and procure capabilities similar to those of the USAF (and the U.S. armed forces more generally) is a matter of some debate.[2]

Subject-matter experts we spoke with for this study gave differing responses to the question of whether the PLAAF is copying U.S. military aerospace; some chose to highlight the ways in which the PLAAF differs from the USAF. "I don't see the PLA as copying the U.S. military much," stated one long-time observer who argued that the overall differences between the two militaries are often overshadowed by some superficial similarities.[3] Another noted that

> the PLAAF doesn't have the space mission the way the USAF does—that's what the PLASSF does. The PLAAF doesn't do rockets—that's the PLARF. They have traditionally focused on air defense of the homeland [which has not had to be a main focus of the USAF as a result of the United States' weak, friendly neighbors and large maritime buffers]. The PLAAF has not focused on strategic bombing the way the USAF has. And the PLA as a whole has a Leninist political commissar structure that has no analogue in the U.S. military.[4]

These are indeed key differences that analysts should bear in mind even when considering possible similarities and instances of imitation or copying.

Operationally, some analysts perceived substantial differences as well. "The PLA may not follow the USAF approach on the employment of UAVs," one analyst averred, noting that the

---

[1] This concept was recently mentioned by Academy of Military Sciences researcher Du Wenlong during an interview that also featured CMC member and PLAAF Commander Ma Xiaotian as well as PLAAF Political Commissar Yu Zhongfu. See "Telling the Air Force's Story Well Is Another Fist for Dealing with 'Air Noise' and 'Sea Noise'" ["讲好空军故事" 是应对 "空闹" "海闹" 的另一只拳头], *People's Daily Online* [人民网], July 22, 2016.

[2] As the term is used here, Chinese efforts to copy the USAF could entail imitating key aspects of the USAF's aerospace development plans; acquiring similar platforms; using airpower in similar ways; or adopting a similar mix of forces (fighters, bombers, ISR platforms), among other elements of military aerospace power development. This does not mean that all aspects of the PLA's aerospace are an exact imitation of the USAF, but rather that China copies some aspects of the USAF (as well as some other militaries) while adapting these to its ends. A comprehensive, definitive assessment of the extent of PLA copying would require access to classified information and more information about technical specifics of certain platforms than the available open-source data permitted. The author thanks Michael Lostumbo for suggesting that he clarify this point.

[3] RAND interview 3, July 2017.

[4] RAND interview 9, July 2017.

PLA use of drones will be different "because they're being used for different reasons and to support different political goals. The PLA would likely operate drones in a very different environment than the United States deployed its UAVs in. The [United States] faced a permissive air environment, whereas the PLA probably would not. The [United States] deployed drones as a way to avoid risking pilots' lives," whereas the PLA would be using drones primarily for strike missions or gray zone coercion against states such as Japan, Taiwan, Vietnam, or India.[5] Similarly, another analyst pointed out that, with regard to rotary-wing aircraft, the PLAA

> only started to field helicopters after about 1991, and these were mostly transport, not attack helicopters. The PLA did not develop its rotary-wing capabilities for combat or medical evacuation [i.e., missions that the United States uses helicopters for] because the Chinese are not doing [CAS or] counter-insurgency. At sea, the PLAN uses their helicopters for antisubmarine warfare, maritime domain awareness, mine counter-measures, and search-and-rescue, [but rarely for forced or opposed-entry boarding of suspect vessels].[6]

Still, one analyst of the Chinese armed forces pointed out, over time and as the mission sets and challenges the PLA faces change, China may choose to copy or innovate solutions in these areas if they become more important to accomplishing the policy goals set for the PLA by the CCP.[7]

Other analysts, while acknowledging these differences, saw similarities worth calling attention to, especially in trend lines related to air power. "There is a lot of imitation across the board," one analyst noted, pointing out that

> while there are some areas where they don't imitate [the United States], in most areas where [the United States] has [a given capability], [the Chinese] want it [too], even to the point of making direct copies. Over time, they're not looking more indigenous or unique—they're looking more and more like [the U.S. military].[8]

This is probably in part because the Chinese are looking to do the types of things the U.S. military does, and also because they have observed the U.S. military and are imitating its successful solutions to the problems of operating over large distances. Another expert echoed this point, noting that "their need for SLOC [sea lines of communication] protection and trade make them more interested in [the United States] because we do all of that. They are likely to become more expeditionary" over time, and this will fuel a continuing interest in learning from the U.S. experience of overseas basing and support of power projection over long distances, including for missions like executing noncombatant evacuation operations.[9] Similarly, to the

---

[5] RAND interview 1, June 2017.

[6] RAND interview 12, August 2017.

[7] The author thanks Cortez A. Cooper, III, for suggesting that he make this point more explicit.

[8] RAND interview 2, July 2017.

[9] RAND interview 7, July 2017.

extent that China adopts an approach to power projection based on aerially refueled stealthy manned bombers, points out one analyst, in the future the PLAAF "will look more like us."[10]

Still other analysts cautioned that the very notion of copying was probably the wrong framework to employ, or saw elements of the notion of copying that they were on guard against. One argued that China "is not copying us, they're studying us—our strengths, our weaknesses, our failures."[11] Another specialist on the PLA echoed this view, arguing that in the air domain,

> they're studying everybody and implementing concepts of operations and [procuring the] hardware [necessary to support these wherever they are] useful to their needs. Lacking experience of their own, they necessarily have to copy a lot. They copy what's relevant to their system and needs. In terms of air operations, they only learn about these from [the United States]; [the United States is] also the only opponent they build against. They don't copy [the United States except] to be able to beat [it].[12]

A third pointed out that the PLA is "not exactly copying but rather absorbing U.S., Russian, and French technologies."[13] A fourth expert noted that the PLA

> looks at how we go to war and plans against that, including how our air force goes to war. The PLA has structured itself to beat that force . . . for example, the J-20 was designed with long legs [i.e., great unrefueled flying range] and a substantial payload capacity so as to fight [the United States] in the Western Pacific. Similarly they've built up lots of 4th generation fighters, ballistic and cruise missiles, AWACS, tankers, and are developing a long-range stealth bomber all designed for war with the [the United States].[14]

A final point that some analysts called attention to is that the PLA's copying clearly is not undertaken without consideration of the missions that China needs to accomplish. The PLA appears to copy foreign militaries when it can find low-cost hardware, organizational, or operational concepts that it can adapt from abroad to solve the operational challenges it confronts. In contrast, when foreign capabilities or organizational practices are irrelevant to Chinese military aerospace problem sets, the PLA either innovates its own solution or ignores the foreign capability.[15]

---

[10] RAND interview 9, July 2017.

[11] RAND interview 6, July 2017.

[12] RAND interview 5, July 2017.

[13] RAND interview 10, July 2017.

[14] RAND interview 7, July 2017.

[15] Interestingly, apart from some military aerospace hardware, experts saw little in the PLA's overall portfolio of capabilities, organization, and operating concepts that it appears to have adopted from the Russian and Ukrainian militaries in recent years. Indeed, some Chinese observers of military aerospace power have been critical of the PLA's continued reliance on Russian and Ukrainian hardware, while others bemoan the fact that the Chinese defense industrial base appears incapable of producing the platforms the PLA needs, referring to the importation of the Russian Su-35, for example, as evidence of a "negative trend in nationalism" (逆流民族主义). See "What Does

In practice, this means that China does not copy some things the U.S. military does. For example, it is only recently that China has sought to acquire greater aerial lift and refueling capabilities.[16] This is because up until a few years ago China did not have any political need to project power over great distances and was under no particular political imperative to be able to do things like carry out a noncombatant evacuation operation in Northern Africa, or surveil and strike targets out beyond 1,000 nautical miles. One interviewee urged further thinking about the PLA's copying, asking the questions:

> Has the PLA accepted the mobility mission to the same extent as the USAF? How do transport guys [i.e., personnel within the PLAAF whose specialization is in the field of air lift and strategic mobility] do in terms of getting promoted to the rank of general officer within the PLAAF? That they do not probably tells you something about the [limits to the] extent to which the PLAAF is modeling itself after the USAF.[17]

Another expert we spoke with argued that

> the Chinese map their own approach to a variety of problems at the strategic, operational, and tactical levels. Sometimes they are pursuing a different goal [than the United States]; at other times they're simply getting there by a different means. These different [goals and] missions lead to different [outcomes in the way China does things.][18]

Similarly, one of the analysts who talked with us pointed out that

> the things that [the PLA doesn't copy from the United States] are probably absent because they don't want these capabilities, not because they can't copy them. For example, some things the U.S. military does are things it undertakes because it is playing an "away game"; an "away game" is not the PLA's thing. They want and expect to operate out of bases in China and in proximity to their logistics and command and control centers. There is no indication that the PLA desires to possess long-range assets for missions other than those relevant for a Western Pacific scenario.[19]

China has likewise not focused much on CAS since it has no recent experience with close-quarters ground combat and appears to anticipate and be oriented more toward fighting a rival air force or navy rather than supporting fielded ground forces engaging an adversary on land.[20]

China's Air Force Still Need to Learn from Russia? Su-35 Complements Domestic Production Capabilities" (中国空军还有什么向俄罗斯学? Su-35 补足国内产能), *Sina Military News* (新浪军事), January 2, 2017.

[16] RAND interview 2, July 2017.

[17] RAND interview 1, June 2017.

[18] RAND interview 12, August 2017.

[19] RAND interview 5, July 2017.

[20] This is not to say that the PLAA is not preparing for ground combat in a Taiwan, North Korea, Vietnam, or India contingency, but rather that there does not appear to be a substantial emphasis on CAS missions and that the primary foci of China's military development efforts have been air, naval, missile, counterspace, and electronic warfare missions. The PLAA, in contrast, has continually seen its predominance eroded since 1978. It is worth noting that

Similarly, "the Chinese plan on operating in areas close to China so haven't needed space much for targeting" and have instead identified it primarily as a domain where the U.S. military's heavy reliance on overhead satellite constellations for positioning, navigation, and timing (PNT), communications, and ISR constitutes a weakness that can be exploited, one expert noted.[21] Dean Cheng of the Heritage Foundation has noted that China's quest for seizing "information dominance" (制信息权) requires it to master the ultimate high ground of outer space.[22] This is not the same, however, as saying that China's thinking about how to use space is entirely imitative of the U.S. approach. "For the foreseeable future, China will not need space the way the [United States] does," one expert commented, noting that China's power projection ambitions are different and require less support from space; at the same time, its strategic risk tolerance levels may be greater, for which reason

> China has not chosen to put money into space-based satellites for ballistic missile early warning [the way the United States has]. Instead, China's space efforts have focused on signals intelligence-gathering and espionage satellites; the use of space to support its economic development and soft power; and the development of space-denial [i.e., antisatellite] capabilities.[23]

Finally, some concepts are difficult to copy simply through observation or theft of technology designs and secrets. As one interviewee commented,

> The kind of logistics needed to support long-term operations is something they can't learn by studying other militaries, but only by actually conducting long-term operations. Similarly, dynamic targeting beyond day three of a warfight is something they can't learn by just watching foreign militaries operate; you can't learn complex air tactics just by watching either. And in terms of hardware, the design of the plane is really the only part of the puzzle of 5th generation air operations that they have; it's not clear that they can do any of the rest, from jointness to distributed [targeting and command and control, etc.].[24]

In short, China appears to copy and adapt aspects of the U.S.'s and other nations' military aerospace doctrines, capabilities, training and exercises, etc. when it is useful to the PLA's efforts to generate effective military power in pursuit of the CCP's goals. Chinese analysts study carefully the experiences, capabilities, and trend lines being pursued by all advanced militaries; assess the reasons for success and failure of those armed forces in their missions; evaluate what

---

the PLAAF reportedly did not fly any missions in the last war China fought (in 1979 against Vietnam). Today, the PLA's CAS mission appears to be primarily centered on the use of short-range ballistic missiles controlled by the PLARF and land-attack cruise missiles controlled by the PLAN; the PLAAF's fighter-bombers do not appear to play a role in executing CAS missions. The author thanks Cortez A. Cooper, III, for insights on this point.

[21] RAND interview 12, August 2017.

[22] Dean Cheng, *The PLA's Interest in Space Dominance—Testimony Before the U.S.–China Economic and Security Review Commission*, February 24, 2015.

[23] RAND interview 12, August 2017.

[24] RAND interview 5, July 2017.

works and what doesn't in seeking to deny and defeat the U.S. military; and whenever possible copy U.S. and other advanced military capabilities, organization, and practices, adapting them to fit Chinese realities.

6. China's Innovations in Military Aerospace

While much of China's military aerospace capability appears to be derivative of and adapted from foreign capabilities and practices, there are important areas of Chinese innovation. One specialist suggested that Chinese defense innovation occurs when China faces "an operational need for which a solution used by other countries is not available to China for whatever reason."[1] As many experts have noted, the PLAAF was stood up with Soviet assistance in the years immediately following the establishment of the People's Republic of China in 1949. In contrast, the PLA space program and what is now the PLARF benefited from substantially greater distance from the Soviets and enjoyed the political protection and resources necessary for innovation. This variation in history helps China's record of innovation in defense aerospace.

One recent study of Chinese defense innovation found that "[t]o date, China's military innovation programs have been more technologically than doctrinally or organizationally focused."[2] Even in the platform domain, innovation was slow to take root. For example, from the mid-1990s through the late 2000s, the PLA's focus generally and the PLAAF's focus more specifically was on acquiring large numbers of improved capabilities, together with a smaller number of niche advanced weapons, an approach one specialist described as prioritizing "numbers . . . not technology."[3] Since approximately 2012, however, one specialist noted, the PLA has shifted its focus "from absorption to original innovation and from asymmetric military competition with the United States to a mix of symmetric and asymmetric competition," although the PLA's internal estimates suggest it will require "another five to ten years of time plus resource support in order to complete the intended transformation."[4]

Specialists have observed a number of specific areas in which China's military innovation has been particularly attention-grabbing. Several noteworthy examples include the PLA's use of missiles; China's conceptualization and applications of military power in space; the reforms of the PLA under Xi Jinping, especially the creation of the PLASSF; and China's improving defense industrial sector.

On the PLA's missile forces, one expert commented that these have been a key element of the PLA's overall innovativeness, noting that

> conventional ballistic missiles with ranges of over 300 miles are a Chinese innovation. This was an innovative solution for the problem of how to attack an adversary when your fighter/bombers would get shot down. Then somebody in

[1] RAND interview 2, July 2017.

[2] Cheung, 2013, p. 39.

[3] RAND interview 12, August 2017.

[4] RAND interview 10, July 2017.

> the PLA realized that maybe they could also use these assets [i.e., long-range conventional ballistic and cruise missiles] against aircraft carriers too.[5]

Another expert commented that the PLARF is

> innovative and creative, not just on hardware but on [force employment concepts]. They're basically going to take any [military operational problem] that's put before them and then see first if it can be solved with a missile. [While] the PLAAF is very noninnovative, the PLASAF is clearly much more culturally innovative. When the PLAAF and the PLAN were overwhelmed and incapable of matching the U.S. military challenge [in the mid-1990s and early 2000s], the Second Artillery [today's PLARF] said in essence "we can help solve this problem."[6]

Some of the most innovative and fast-moving areas of Chinese military aerospace involve the PLA's investments in hypersonic glide vehicles[7] and research into the use of quantum entanglement for satellite communications encryption.[8] Other research, while perhaps less innovative, nonetheless helps to redress the PLA's weaknesses in the realm of antisubmarine warfare, which it appears to be focusing on through the use of a combination of manned and unmanned platforms for aerial surveillance and detection of foreign submarines.[9] These aerial detection capabilities are likely to be employed in tandem with a subsurface sound sensor network that the PLA is reportedly looking at building across the South China Sea, a network that at least some observers believe could be employed for submarine detection.[10] Collectively, these developments reflect attempts to develop workarounds for U.S. and allied advanced ballistic missile defenses and to counter U.S. advantages in electronic and submarine warfare.

China has also been extremely active in developing its outer space warfare capabilities. Another interviewee commented that, owing to the remote and obscure nature of the space domain,

> it's hard to tell what China has copied or reverse-engineered as opposed to innovated [in its space program]. They probably got a lot of their space technology from Russia, but it's hard to say exactly what. They certainly stole some from the U.S., including liquid oxygen technology and radiation-hardened integrated circuitry. The PLA has clearly determined that they need a set of space capabilities on par with the [United States]. They are clearly thinking about war-fighting, not just deterrence, since you don't extend your ability to strike satellites to the geosynchronous Earth orbit range unless you're planning to fight.

---

[5] RAND interview 2, July 2017.

[6] RAND interview 5, July 2017.

[7] Erika Solem and Karen Montague, "Updated—Chinese Hypersonic Weapons Development," *Jamestown Foundation China Brief*, Vol. 16, No. 7, April 21, 2016.

[8] "Chinese Quantum Satellite Sends 'Unbreakable' Code," Reuters, August 10, 2017.

[9] Mike Yeo, "China Deploys New Anti-Submarine Aircraft to Fringes of South China Sea," *Defense News*, June 22, 2017.

[10] James Griffiths, "Beijing Plans Underwater Observation System in South China Sea," CNN, May 30, 2017.

> And this is driving their more innovative program investments in areas such as quantum communications, hypersonic glide vehicles, pulsar navigation for satellites, and the quest to build an EM [electro-magnetic thruster] drive [for spacecraft propulsion].[11]

The reforms of the PLA from late 2015 onward clearly reflected the growing importance of military aerospace capabilities, prioritizing space capabilities and promoting the PLARF from a branch of the PLAA to an independent service. Additionally, the reforms stood up the PLASSF as a new and centralized body to exert control over key aerospace capabilities, including space, cyber, electronic warfare, and information warfare assets.[12] Under the new arrangement, these capabilities are centralized so as to produce "'integrated reconnaissance, attack, and defense' (侦攻防一体化)."[13] Elsa Kania argues that

> [a]lthough the PLA has typically been characterized as an organization resistant to change, the [PLASSF], as a specialized, technical force with a mandate and an identity that center upon innovation, might be better poised to advance defense innovation than the PLA at large.[14]

John Costello, a leading specialist on the PLASSF, points out that compared with the United States,

> [t]he [PLA] is taking an approach quite different from that of the U.S. military, in which each service builds its information warfare forces and then integrates them at the joint combatant commands. Instead, the PLA is creating an entire military service branch dedicated to information warfare, comprising space, cyber, and electronic warfare units that will form a core around which the other services will operate in the information domain.[15]

This is a new and innovative approach to managing information warfare; its effectiveness will be tested over time.

Finally, the reforms have created a new CMC Armament Development Department to replace the GAD with the aim of further enhancing defense industrial efficiency.[16] This body is

---

[11] RAND interview 11, July 2017.

[12] According to PLA commentator Yin Zhuo, the PLASSF's primary mission is to "support battlefield operations, allowing our military to achieve localized superiority in aviation, space, network and electromagnetic battlefields." See Li Gangshe [李刚摄], "Expert: Strategic Support Force Will Link the Entire Operational Process, Be Key to Victory" [专家：战略支援部队将贯穿作战全过程 是致胜关键], *People's Daily Online* [人民网], January 5, 2016.

[13] John Costello, "The Strategic Support Force: China's Information Warfare Service," *Jamestown Foundation China Brief*, Vol. 16, No. 13, February 8, 2016.

[14] Elsa Kania, "China's Strategic Support Force: A Force for Innovation?" *The Diplomat*, February 18, 2017.

[15] John Costello, "China Finally Centralizes Its Space, Cyber, and Information Warfare Forces," *The Diplomat*, January 20, 2016. The elevation of U.S. Cyber Command to the status of a unified combatant command on August 18, 2017, would appear to reduce Chinese and U.S. differences in this area.

[16] Details on the impact of the reforms on China's defense industrial bureaucracy can be found in Tai Ming Cheung, Thomas Mahnken, Deborah Seligsohn, Kevin Pollpeter, Eric Anderson, and Fan Yang, *Planning for Innovation:*

mirrored in each service by an Armaments Development Department intended to ensure that research and development, planning, and procurement are more smoothly and effectively synchronized and executed. Analysts agreed that in some important ways the Chinese defense industrial sector has already outgrown its previous image as a stodgy and inflexible jobs program incapable of developing relevant capabilities; the reforms may further advance the Chinese defense industrial sector's ability to produce the weapons systems and platforms that the PLA seeks. "They are more agile than the [United States] in terms of getting platforms to the field quickly," noted one observer, arguing that the PLA as a whole is increasingly "innovative at thinking about how to go to war, how to employ [force], and [identifying] what [capabilities] to develop."[17] Another specialist agreed, saying that

> the most innovative things they've done are in transforming the structure and process of defense research and development so as to move up the time from the start of research to the fielding of an initial operational capacity.[18]

While the large defense industrial conglomerates are probably still less efficient and innovative than their Western counterparts, the PLA's inputs into the system of capabilities development have definitely accelerated the process of fielding relevant weapons systems. China's narrow focus on a single adversary (the United States) also helps efficiency, and the Chinese process for approving designs and moving forward to prototyping, testing and evaluation, and initial production appears to pose fewer obstacles to developers. These factors result in relevant military aerospace platforms being fielded faster than the two- to three-decade development cycles of Western defense aerospace firms. Chinese aerospace firms, of course, operate in a different legal and sociopolitical context, where lawsuits, oversight, and popular views of defense spending are lesser constraints than in the United States or Western Europe.

Chinese military aerospace is innovative largely in areas isolated from the historical deadweight of ties to the Soviet model of military organization and capabilities development and in evolving asymmetric solutions focused on the delivery of munitions where manned fighter-bombers would be unlikely to survive. It has also proven innovative in designing ways to attack perceived U.S. vulnerabilities in space, cyberspace, and the information realm. China has also shown some innovation in leveraging the relative immunity of the Chinese defense sector from public oversight and accountability, which allows it to tolerate more risk and hurry specific, focused capabilities into the field. The extent to which these narrow areas of innovativeness can be expanded to the broader PLA is unclear. However, a gradually improving set of conventional capabilities (largely copied from the West), paired with a more narrow portfolio of highly innovative asymmetric capabilities focused against perceived U.S. vulnerabilities, may solve the

---

Understanding China's Plans for Technical, Energy, Industrial, and Defense Development, Washington, D.C.: U.S.–China Economic and Security Review Commission, 2016, pp. 19–24.

[17] RAND interview 7, July 2017.

[18] RAND interview 4, July 2017.

problems China faces in confronting a more technologically sophisticated adversary—an adversary that is hampered by its need to project power far from its home territory in necessarily limited numbers.

7. Conclusions and Implications

As the analysis presented thus far makes clear, the PLA has increasingly prioritized the development of military aerospace power in recent years to achieve the goals identified by the CCP leadership, even if these goals require defeating the U.S. military on the battlefield. As the PLA seeks to deny U.S. advantages and project power in the face of U.S. technological superiority, it has copied when it can and innovated where it must. China has studied the importance of U.S. dominance in military aerospace power, including the air domain and information dominance in recent U.S. military actions in Europe and the Middle East. However, China has not just copied U.S. operations but innovated in its approach to projecting power in the face of U.S. conventional superiority in the Asia-Pacific. China's power projection is largely built around the development of precision-strike ballistic and cruise missiles, supplemented by a dense web of SAMs and advanced fighters (which appear increasingly likely to be supplemented in the not-too-distant future by deck-borne, fixed-wing aviation).

It is critical that the USAF understand the advances that China is making in specific domains related to ISR, strategic and tactical lift, and strike platforms and assets, as well as power projection in and through space and against space-based satellite architectures. Analyses of Chinese aerospace development can also help the USAF (and the U.S. military and national security establishment more broadly) to identify possible or probable vectors along which the PLA is likely to target U.S. interests in any armed clash. Such knowledge is also useful during times of peace to prevent military-to-military contact that might unwittingly help the PLA further advance its capabilities.[1]

I emphasize that while much public attention focuses on China's increasingly advanced fixed- and rotary-wing hardware and munitions, a wide range of other PLA investments and changes, including in the realms of doctrine, organization, training, manpower, logistics, procurement, and facilities, are critical to the PLA's ability to employ aerospace power effectively; these also warrant close attention. For example, China's recent reorganization of the PLA's C2 functions is clearly intended to achieve the goal of generating integrated, joint operations under the tight political control of the CCP leadership so as to "fight and win wars." Improvements in PLAAF training are intended to prepare the PLA for actual combat operations under real-world conditions.[2] The PLAAF's expanded use of flights in spaces around both Japan

---

[1] China's interest in military-to-military ties with the United States and the Congressional legislation and policy constraints on such contacts are reviewed in Scott W. Harold, "Expanding Contacts to Enhance Durability: A Strategy for Improving US-China Military-to-Military Relations," *Asia Policy*, No. 16, July 2013, pp. 103–137. For example, China has sought to acquire insights into U.S. aircraft carrier–based fixed-wing aviation, although the United States ultimately declined to engage with the PLA on this matter.

[2] Lyle J. Morris and Eric Heginbotham, *From Theory to Practice: People's Liberation Army Air Force Aviation Training at the Operational Unit,* Santa Monica, Calif.: RAND Corporation, RR-1415-AF, 2016; Lyle J. Morris,

and Taiwan are intended both to facilitate familiarity with possible combat arenas and also to blur some of the indicators and warnings that Tokyo, Taipei, and Washington would use to detect an incipient attack.[3] Perhaps even more importantly, the ultimate goal of such exercises is to give the PLAAF a credible capacity to "deter U.S. intervention in a regional conflict"; they are also valuable insofar as they elicit information about Taiwan's defenses, response times, and reaction patterns.[4]

The PLAAF's expanded flight training has been paired with expanded use of naval training and exercises in the waters within the first island chain and out to the Western Pacific, further highlighting the importance of tracking overall PLA developments even as particular attention is paid to China's military aerospace, since such operations are ultimately intended to be joint and integrated.

Improvements in the PLA's defense industrial base—including its increasing integration with the civilian Chinese economy, its growing sophistication, and its adept workarounds on foreign technology controls so as to procure advanced technologies that it cannot produce indigenously—are also critical enablers of the PLA's advancing military aerospace capabilities.[5] Particularly worth noting are China's expanding exports of airframes, air defenses, and missiles, giving it an opportunity to lower per unit costs, recoup investments, and potentially test its hardware in foreign conflicts without actually having to fight.[6]

Finally, improvements to Chinese air basing and the deployment of fixed-wing air assets and SAMs to the islands that China controls in the South China Sea enhance the PLA's ability to

---

"China's Air Force Freeing Fighters from Scripted Tactics," *Aviation Week & Space Technology*, October 14, 2016; Lyle J. Morris and Eric Heginbotham, "China's PLAAF Pilot Training Program Undergoes Major Overhaul," *National Interest*, October 27, 2016.

[3] Jesse Johnson, "Chinese Air Force Conducts 'Several' Long-Range Drills Near Japan as Military Tells Tokyo to 'Get Used to It,'" *Japan Times*, July 16, 2017; "Taiwan Says Chinese Aircraft Fly Around Island in Weekend of Drills," Reuters, August 13, 2017.

[4] Nathan Beauchamp-Mustafaga, Derek Grossman, and Logan Ma, "Chinese Bomber Flights Around Taiwan: For What Purpose?" *War on the Rocks*, September 13, 2017.

[5] On China's quest to improve its overall aerospace production capacity and build its own indigenous civilian airframes, see Roger Cliff, Chad J. R. Ohlandt, and David Yang, *Ready for Takeoff: China's Advancing Aerospace Industry*, Santa Monica, Calif.: RAND Corporation, MG-1100-UCESRC, 2011; Keith Crane, Jill Luoto, Scott W. Harold, David Yang, Samuel K. Berkowitz, and Xiao Wang, *The Effectiveness of China's Industrial Policies in Commercial Aviation Manufacturing*, Santa Monica, Calif.: RAND Corporation, RR-245, 2014; Chad Ohlandt, *Chinese Investment in U.S. Aviation*, Santa Monica, Calif.: RAND Corporation, RR-1755-USCC, 2017. On China's increasing embrace of the private sector as having a role in defense industrial development under the Xi Jinping administration, see "Xi to Head Civil-Military Integration Body," *Global Times*, January 23, 2017. On China's expanding defense exports, including of aerospace technologies (especially drones and fixed-wing airframes), see Franz Stefan-Gady, "China Scores Biggest Military Export Order for Killer Drones," *The Diplomat*, March 2, 2017; Allan Nixon, "China's Growing Arms Sales to Latin America," *The Diplomat*, August 24, 2016.

[6] See Appendixes A, B, and C. All information generated from Stockholm International Peace Research Institute Arms Transfer database; search terms used were "Chinese Transfers of Air Defense Systems, Combat Aircraft, and Missiles to World, 2010–2017."

engage for deterrence, gray-zone coercion, and wartime operations.[7] China's willingness to use unmanned systems to engage in gray-zone coercion is worth monitoring, as it is unclear exactly how nations should regard the intrusion of unmanned platforms or how China would respond to having one of its drones shot down.[8]

In conclusion, the PLA is focused not merely on competing with the United States or other nations as a goal in and of itself, but instead on competing as a means to achieving the policy outcomes identified by the CCP—deterring U.S. intervention and defeating the U.S. military if the United States and China do come into open conflict. The PLA does not blindly copy the USAF (or the U.S. military or other foreign militaries more generally) but instead studies foreign military experiences to learn what works and what does not, what can be adopted and adapted to serve China's goals, and what should be ignored or discarded as irrelevant to the missions set before the PLA by the CCP. When the PLA finds problems that it cannot solve through copying, it has proven sufficiently resourceful to develop innovative solutions to its operational challenges, creating entirely new capabilities or operating in new and creative ways to frustrate the ability of the United States and other advanced militaries to operate in proximity to Chinese shores.

It is fair to return to a point made repeatedly in the interviews with PLA subject-matter experts, who pointed out that the main drivers for the development of the Chinese armed forces are the requirements that the PLA faces in order to achieve the goals of the CCP. The directive from Xi Jinping—that the PLA prepare to "fight and win wars"—shows that today's Chinese armed forces must be prepared to find a way to defeat, not merely compete with, the United States; in so doing, copying and innovation are both valid pathways to the extent that they serve this goal.

---

[7] Elizabeth Shim, "China Deploys Air Force on Disputed South China Sea Island," United Press International, October 20, 2016; "A Look at China's SAM Shelters in the Spratlys," *Center for Strategic and International Studies Asia Maritime Transparency Initiative*, February 23, 2017.

[8] "Drone Joins Four Chinese Ships in Latest Senkaku Intrusion," *Kyodo*, May 18, 2017; Ankit Panda, "Japan to Shoot Down Foreign Drones," *The Diplomat*, October 22, 2013; "Former Chinese Commander Warns of War If Japan Shoots Down Drone," *Bloomberg*, November 4, 2013.

Appendix A. Transfers of Air Defense Systems, China to World (Final Deliveries and New Orders, 2010 to 2017)

Recipient	Number Ordered	Weapons Designation	Weapon Description	Year of Order	Year(s) of Delivery	Number Delivered	Comments
Bangladesh	2	FM-90	SAM system	2010	2011	2	
	2	FM-90	SAM system	2015	2016	2	
Cameroon	12	GDF 35-mm	Antiaircraft gun	2012	2013	12	Type-90 (PG-99) version
Ethiopia	1	HQ-64	SAM system	2012	2013	1	
Indonesia	8	TD-2000B	Air defense system	2008	2012–2013	8	Part of $35 million deal
	4	GDF 35-mm	Antiaircraft gun	2015	2016	4	Type-90 version
Morocco	6	Sky Shield	Air defense system	2010	2011–2012	6	
Myanmar	4	KS-1A	SAM system	2013	2015–2016	4	
Pakistan	20	GDF 35-mm	Antiaircraft gun	2011	2012	20	Type-90 version
	10	FM-90	SAM system	2013	2014–2016	10	
	3	LY-80 SAMS	SAM system	2014	2015–2016	3	$599 million deal
Sudan	2	FB-6	Mobile SAM system	2015	2016	2	
Tanzania	1	FB-6	Mobile SAM system	2012	2013	1	
Thailand	1	KS-1A	SAM system	2016	2016	1	KS-1C version
Turkmenistan	1	FM-90	SAM system	2015	2016	1	
	1	HQ-9	SAM system	2015	2016	1	
	1	KS-1A	SAM system	2015	2016	1	Probably KS-1C version

SOURCE: Stockholm International Peace Research Institute, SIPRI Arms Transfer database.
NOTE: The “Numbers Delivered” and the “Year(s) of Deliveries” columns refer to all deliveries since the beginning of the contract. The “Comments” column includes publicly reported information on the value of the deal.

Appendix B: Transfers of Combat Aircraft, China to World (Final Deliveries and New Orders, 2010 to 2017)

Recipient	Number Ordered	Weapons Designation	Weapon Description	Year of Order	Year(s) of Delivery	Number Delivered	Comments
Egypt	18	ASN-209	UAV	2010	2012–2014	18	
Pakistan	50	JF-17 Thunder/FC-1		1999	2007–2013	50	JF-17 Block-1 version; developed for Pakistan, including component production and assembly in Pakistan; including 8 mainly for testing; first 42 of production version ordered in 2009 for $800 million
	27	K-8 Karakorum-8	Trainer/ combat aircraft	2005	2007–2010	27	K-8P version
	20	CH-3	UAV/UCAV	2009	2013–2016	20	
	50	JF-17 Thunder/FC-1	FGA aircraft	2011	2015–2017	20	JF-17 Block-2 version
	50	JF-17 Thunder/FC-1	FGA aircraft	2012	Possibly from 2019		JF-17 Block-3 version
Bangladesh	16	F-7MG	Fighter aircraft	2011	2012–2013	16	F-7BGI version
	9	K-8 Karakorum-8	Trainer/ combat aircraft	2013	2014–2015	9	K-8W version
	11	BT-6/PT-6	Trainer aircraft	2015	2016–2017	11	
Bolivia	6	K-8 Karakorum-8	Trainer/ combat aircraft	2009	2011	6	$58 million deal; K-8WB version
	6	AS365/AS565 Panther	Helicopter	2012	2014	6	$108–113 million deal; H-425 (Z-9) version
Cambodia	12	AS365/AS565 Panther	Helicopter	2012	2013	12	$195 million deal (Chinese loan); including four armed version
	2	MA60	Transport aircraft	2012	2012	2	
Cameroon	2	MA60	Transport aircraft	2011	2013	2	
	4	Z-9WZ	Combat helicopter	2013	2014	4	Z-9WE version
Djibouti	1	MA60	Transport aircraft	2013	2014	1	

Recipient	Number Ordered	Weapons Designation	Weapon Description	Year of Order	Year(s) of Delivery	Number Delivered	Comments
	2	Y-12	Light transport aircraft	2015	2016	2	Y-12E version
Egypt	40	K-8 Karakorum-8	Trainer/ combat aircraft	2004	2007–2010	40	K-8E version; assembled from kits in Egypt
	3	Wing Loong-1	UAV/UCAV	2016	2017	3	
Ghana	4	AS365/AS565 Panther	Helicopter	2014	2015	4	Z-9 version
Indonesia	4	Wing Loong-1	UAV/UCAV	2017			
Iraq	4	CH-4	UAV/UCAV	2014	2015	4	Armed CH-4B version
Jordan	2	CH-4	UAV/UCAV	2015	2016	2	CH-4B version
Kazakhstan	2	Wing Loong-1	UAV/UCAV	2015	2016	2	
Kenya	4	AS365/AS565 Panther	Helicopter	2009	2010	4	Z-9WA armed version
	5	AS365/AS565 Panther	Helicopter	2012	2014–2015	5	Z-9WA armed version
Laos	2	MA60	Transport aircraft	2010	2012	2	MA-600 version
	2	MA60	Transport aircraft	2012	2013	2	MA-600 version
	9	LE-500 Little Eagle	Light aircraft	2015	2015–2017	9	
Mali	2	Y-12	Light transport aircraft	2016	2017	2	Y-12E version
Myanmar	50	K-8 Karakorum-8	Trainer/ combat aircraft	2009	2011–2016	24	Number could be 60 or 72; assembled in Myanmar
	12	CH-3	UAV/UCAV	2013	2014–2015	12	
	16	JF-17 Thunder/FC-1	FGA aircraft	2015	2017	4	Possibly from Pakistani production line
	2	Y-8	Transport aircraft	2015	2016	2	Y-8F-200W version
Namibia	2	AS365/AS565 Panther	Helicopter	2009	2012	2	H-425 (Z-9) version
Nigeria	15	F-7M Airguard	Fighter aircraft	2005	2010	15	$251 million deal; F-7NI version, including 3 FT-7NI version
	5	CH-3	UAV/UCAV	2014	2014	5	Armed UCAV
Pakistan	6	AS565S Panther	Antisubmarine helicopter	2005	2009–2010	6	Z-9EC version
	4	ZDK-03	Airborne early warning and control aircraft	2008	2011–2014	4	$278 million deal; designated KE-03 in Pakistan

Recipient	Number Ordered	Weapons Designation	Weapon Description	Year of Order	Year(s) of Delivery	Number Delivered	Comments
	4	F-7A/J-7	Fighter aircraft	2010	2010	4	Secondhand; FT-7A version; aid
	3	WZ-10	Combat helicopter	2014	2015	3	
	2	Wing Loong-1	UAV/UCAV	2015	2015	2	
Saudi Arabia	2	CH-4	UAV/UCAV	2014	2015	2	
		Wing Loong-2	UAV/UCAV	2017	2017	5	
Seychelles	2	Y-12	Light transport aircraft	2011	2011	2	Aid; including for maritime patrol
Slovakia	1	A319	Transport aircraft	2016	2017	1	Secondhand (from Hong Kong–based civilian operator); A319-100 version
Sri Lanka	2	Y-12	Light transport aircraft	2009	2010	2	
Sudan	6	FTC-200	Trainer/ combat aircraft	2015	2017	3	
Tanzania	14	F-7MG	Fighter aircraft	2008	2009–2012	14	F-7TN version; including 2 FT-7TN version
	6	K-8 Karakorum-8	Trainer/ combat aircraft	2010	2011–2012	6	
Turkmeni-stan	2	CH-3	UAV/UCAV	2015	2016	2	
	2	WJ-600	UAV/UCAV	2015	2016	2	
United Arab Emirates	5	Wing Loong-1	UAV/UCAV	2011	2013–2014	5	
	5	Wing Loong-2	UAV/UCAV	2017	2017	5	
Venezuela	18	K-8 Karakorum-8	Trainer/ combat aircraft	2008	2010	18	
	8	Y-8	Transport aircraft	2011	2012–2014	8	Y-8F-200W version
	9	K-8 Karakorum-8	Trainer/ combat aircraft	2014	2016		9K-8W or K-8VV version
Zambia	8	K-8 Karakorum-8	K-8 Karakorum-8	2010	2012	8	
	7	AS365/AS565 Panther	Helicopter	2011	2012	7	$105 million deal
	6	L-15	Trainer/ combat aircraft	2014	2016–2017	6	L-15AFT version

SOURCE: Stockholm International Peace Research Institute, SIPRI Arms Transfer database.
NOTE: The "Numbers Delivered" and the "Year(s) of Deliveries" columns refer to all deliveries since the beginning of the contract. The "Comments" column includes publicly reported information on the value of the deal. FGA = fighter/ground attack aircraft; UCAV = unmanned combat air vehicle.

Appendix C: Transfers of Missiles, China to World (Final Deliveries and New Orders, 2010 to 2017)

Recipient	Number Ordered	Weapons Designation	Weapon Description	Year of Order	Year(s) of Delivery	Number Delivered	Comments
Iran	380	C-802/ CSS-N-8	ASCM	1992	1994–2012	380	Including Hudong (Thondor), new-produced and modernized Combattante-2 (Kaman) FAC and coast defense systems; possibly including air-launched version; Iranian designation Tondar or Noor; status since 2010 uncertain because of United Nations arms embargo
	260	FL-6	ASCM	1998	1999–2015	260	Developed or copied by China from Italian Sea Killer (Marte-2) antiship missile supplied by Iran to China; Iranian designation Fajr-e Darya; including for SH-3D helicopters; status since 2010 uncertain because of United Nations arms embargo
	150	TL-10/FL-8	ASCM	2002	2004–2015	150	TL-10A and possibly TL-10B version; status 2010 uncertain because of United Nations arms embargo
	50	C-704	ASCM	2003	2010–2010	50	Developed for Iran; Iranian designation Nasr-1
	50	C-801	ASCM	2004	2006–2015	50	Iranian designation Kosar and/or Sagheb; including submarine-launched version; status since 2010 uncertain because of United Nations arms embargo
	650	QW-11	Portable SAM	2005	2006–2015	650	Iranian designation Misagh-2; status since 2010 uncertain because of United Nations arms embargo
Pakistan		Red Arrow-8	Antitank missile	1989	1990–2017	23,850	Pakistani designation Baktar Shikan
		QW-1 Vanguard	Portable SAM	1993	1994–2017	2,150	Pakistani designation Anza-2

Recipient	Number Ordered	Weapons Designation	Weapon Description	Year of Order	Year(s) of Delivery	Number Delivered	Comments
Turkey	200	B-611		1999	2002–2012	200	Turkish designation Yildirim or Project-J or J-600T; possibly including B-611M version
Algeria	50	C-802/ CSS-N-8	ASCM	2012	2015–2016	50	For C-28A frigates
	50	FM-90	SAM	2012	2015–2016	50	For C-28A frigates
Bangladesh	75	FM-90	SAM	2010	2011	75	
	16	C-704	ASCM	2011	2012–2013	16	For LPC-1 corvettes
	30	C-802/ CSS-N-8	ASCM	2012	2014	30	For Type-053 or Type-510 (Jianghu) frigates
	15	C-802/ CSS-N-8	ASCM	2012	2015	15	For Type-056 (Shadhinata) frigates
	50	HHQ-10	SAM	2012	2015	50	For Type-056 (Shadhinata) frigates
	60	Yu-4 553-mm	Antiship torpedo	2013	2016	60	Probably secondhand; for Type-035G submarines; designation uncertain
	100	FN-6	Portable SAM	2015			FN-16 version; possibly including assembly in Bangladesh
	50	HHQ-10	SAM	2015			For Type-056 (Shadhinata) frigates
Cameroon	50	FN-6	Portable SAM	2012	2013	50	FN-16 version
	50	Red Arrow-8	Antitank missile	2013	2014	50	For Z-9 combat helicopters
Ethiopia	75	PL-11/FD-60	Beyond visual range air-to-air missile	2012	2013	75	For HQ-64 SAM system
Indonesia	200	QW-3	Portable SAM	2008	2012–2013	200	For TD-2000B air defense system
	15	QW-3	Portable SAM	2009	2010	15	
	100	C-705	ASCM	2011	2014–2017	66	For KCR-40 and KCR-60 FAC; including assembly from kits or production in Indonesia from 2017/2018
	25	C-802/ CSS-N-8	ASCM	2011	2012–2016	25	For PB-57 (Todak) FAC and Van Speyk (Ahmad Yani) frigates
Iraq	20	AR-1	ASCM	2014	2015	20	For CH-4B UAV
	20	FT-9	Guided bomb	2014	2015	20	For CH-4B UAV
Jordan	20	QW-2	Portable SAM	2014	2014	20	Designation uncertain

Recipient	Number Ordered	Weapons Designation	Weapon Description	Year of Order	Year(s) of Delivery	Number Delivered	Comments
Morocco	75	PL-9	Short-range air-to-air missile	2010	2011–2012	75	For Sky Shield air defense systems
	50	Red Arrow-8	Antitank missile	2010	2010	50	
Myanmar	20	C-802/ CSS-N-8	ASCM	2009	2013–2014	20	For FAC-491
	30	C-802/ CSS-N-8	ASCM	2011	2012	30	For Type-053 (Jianghu-2) frigates
	10	C-802/ CSS-N-8	ASCM	2011	2016	10	For Tabinshwehti corvette produced in Myanmar
	25	HY-2/ SY-1A/ CSS-N-2	ASCM	2011	2012	25	Probably secondhand; for Type-053 (Jianghu-2) frigates
	200	KS-1A	SAM	2013	2015–2016	200	
Nigeria	20	PL-9	Short-range air-to-air missile	2005	2010	20	$20 million deal (part of $32 million deal that included other armaments); PL-9C version; for F-7NI combat aircraft
	30	AR-1	SAM	2014	2014	30	
Pakistan	70	C-802/ CSS-N-8	ASCM	2015	2009–2013	70	For Jiangwei (F-22P) frigates
	100	R-440 Crotale	SAM	2005	2009–2013	100	For Jiangwei (F-22P) frigates; HQ-7 (FM-80) version
	600	PL-12/SD-10	Beyond visual range air-to-air missile	2006	2010–2017	375	For JF-17 and possibly modernized Mirage-3/5 combat aircraft
	1,000	PL-5E	Short-range air-to-air missile	2006	2009–2017	760	For JF-17 combat aircraft; PL-5E-II version
	100	C-802/ CSS-N-8	ASCM	2008	2012–2017	60	For JF-17 combat aircraft
	750	LS-3	Guided bomb	2008	2010–2017	575	For JF-17 combat aircraft
	1,000	LS-6-500	Guided bomb	2008	2010–2017	625	For JF-17 combat aircraft
	750	LT-2	Guided bomb	2008	2010–2017	550	For JF-17 combat aircraft
	50	FN-6	Portable SAM	2009	2010	50	
	30	C-802/ CSS-N-8	ASCM	2010	2012–2014	30	For Azmat FAC
	50	CM-400AKG	ASCM	2010	2012–2016	50	For JF-17 combat aircraft

Recipient	Number Ordered	Weapons Designation	Weapon Description	Year of Order	Year(s) of Delivery	Number Delivered	Comments
	100	LD-10	ARM	2011	2014–2017	100	
	30	C-802/ CSS-N-8	ASCM	2013	2017	15	For Azmat FAC
	400	FM-90	SAM	2013	2014–2016	400	
	300	LY-80	SAM	2014	2015–2016	300	
	80	C-802/ CSS-N-8	ASCM	2015			For 6 Type-041 (S-20) submarines
	100	SET-65 Yenot-2	Antisubmarine weapon torpedo	2015			Yu-3 version; for Type-041 submarines
	100	Yu-4 533-mm	Antisubmarine torpedo	2015			For 6 Type-041 submarines
Peru	15	FN-6	Portable SAM	2009	2010	15	$1.1 million deal
Qatar	12	BP-12A	Surface-to-surface missile	2016	2017	12	For SY-400 multiple rocket launcher
South Africa	50	FN-6	Portable SAM	2015	2016	50	
South Sudan	1,200	Red Arrow-73	Antitank missile	2013	2014	1,200	$14.5 million deal (including 100 launchers); Red Arrow-73D version
Sudan	450	Red Arrow-8	Antitank missile	2009	2009–2012	450	
	100	FN-6	Portable SAM	2015	2016	100	
Syria	500	Red Arrow-73	Antitank missile	2013	2014	500	Red Arrow-73D version
Tanzania	50	FN-6	Portable SAM	2012	2013	50	For FB-6 SAM system
Thailand	60	C-802/ CSS-N-8	ASCM	2007	2009–2014	60	For modernized Chao Phraya frigates
		WS-3A 300-mm	Guided rocket	2012	2016	25	For WS-1 (DTi-1) multiple rocket launcher
	50	KS-1A	SAM	2016	2016	50	KS-1C version
		C-708UNA	ASCM	2017			For S26T submarine
		YU-8 533-mm	Antiship/ antisubmarine weapon torpedo	2017			For S26T submarine; designation uncertain (reported as “torpedo”)
Turkmenistan	10	AR-1	ASCM	2015	2016	10	For CH-3 UAV/UCAV
	10	CS-502KG	ASCM	2015	2016	10	For WJ-600 UAV/UCAV
	40	FM-90	SAM	2015	2016	40	
	75	HQ-9	SAM	2015	2016	75	
	50	KS-1A	SAM	2015	2016	50	Probably KS-1C version

Recipient	Number Ordered	Weapons Designation	Weapon Description	Year of Order	Year(s) of Delivery	Number Delivered	Comments
Venezuela	100	PL-5E	Short-range air-to-air missile	2008	2010	100	For K-8 trainer/combat aircraft
	250	Red Arrow-73	Antitank missile	2012	2015	250	Red Arrow-73D version for VN-18 IFV
		C-802/ CSS-N-8	ASCM	2017			
Zambia	50	LS-6-500	Guided bomb	2014	2016–2017	50	For L-15 combat/trainer aircraft
	40	PL-5E	Short-range air-to-air missile	2014	2016–2017	40	
	50	YJ-9E	ASCM	2014	2016–2017	50	For L-15 combat/trainer aircraft.

SOURCE: Stockholm International Peace Research Institute, SIPRI Arms Transfer database
NOTE: The "Numbers Delivered" and the "Year(s) of Deliveries" columns refer to all deliveries since the beginning of the contract. The "Comments" column includes publicly reported information on the value of the deal. ASCM = antiship cruise missile.

References

Allen, Kenneth W., and Eric A. McVadon, *China's Foreign Military Relations*, Washington, D.C.: Henry L. Stimson Center, 1999.

Allen, Kenneth W., Glenn Krumel, and Jonathan D. Pollack, *China's Air Force Enters the 21st Century*, Santa Monica, Calif.: RAND Corporation, MR-580-AF, 1995. As of June 21, 2018: https://www.rand.org/pubs/monograph_reports/MR580.html

An Ti [安替], "CMC's New Weapons and Armaments Administration Regulations Advance Modernization of Chinese Military Equipment" [军委武器装备管理新条例推动中国军备现代化], *People's Daily Online* [人民网], January 21, 2003.

Axe, David, "Go Ahead China—Copy Our Crappiest Warplane," *Daily Beast*, June 5, 2015.

Ayyadi, Islam, and Mohammad Kamal, "China-Israel Arms Trade and Cooperation: History and Policy Implications," *Asian Affairs*, Vol. 47, No. 2, 2016, pp. 260–273.

Beauchamp-Mustafaga, Nathan, Derek Grossman, and Logan Ma, "Chinese Bomber Flights Around Taiwan: For What Purpose?" *War on the Rocks*, September 13, 2017.

Berntsen, Gary, and Ralph Pezzullo, *Jawbreaker: The Attack on Bin Laden and al Qaeda—A Personal Account by the CIA's Key Field Commander*, New York, N.Y.: Crown Publishing, 2005.

Blasko, Dennis, "Recent Developments in the Chinese Army's Helicopter Force," *Jamestown Foundation China Brief*, Vol. 17, No. 8, June 9, 2017.

Campbell, Kurt, and Robert Blackwill, *Xi Jinping on the Global Stage: Chinese Foreign Policy Under a Powerful but Exposed Leader—Council Special Report No. 74*, New York: Council on Foreign Relations, February 2016.

Chase, Michael S., Jeffrey Engstrom, Tai Ming Cheung, Kristen A. Gunness, Scott Warren Harold, Susan Puska, and Samuel K. Berkowitz, *China's Incomplete Military Transformation: Understanding the Weaknesses of the PLA*, Santa Monica, Calif.: RAND Corporation, RR-893-USCC, 2015. As of June 21, 2018: https://www.rand.org/pubs/research_reports/RR893.html

Chase, Michael S., Kristen A. Gunness, Lyle J. Morris, Samuel K. Berkowitz, and Benjamin S. Purser III, *Emerging Trends in China's Development of Unmanned Systems*, Santa Monica, Calif.: RAND Corporation, RR-990-OSD, 2015. As of June 21, 2018: https://www.rand.org/pubs/research_reports/RR990.html

Chase, Michael S., and Cristina L. Garafola, "China's Search for a Strategic Air Force," *Journal of Strategic Studies*, Vol. 39, No. 1, 2015, pp. 1–26.

Chase, Michael S., Cristina L. Garafola, and Nathan Beauchamp-Mustafaga, "Chinese Perceptions of and Responses to US Conventional Military Power," *Asian Security*, 2017, pp. 1–19.

Cheng, Dean, *The PLA's Interest in Space Dominance—Testimony Before the U.S.– China Economic and Security Review Commission*, February 24, 2015.

Cheung, Tai Ming, *Fortifying China: The Struggle to Build a Modern Defense Economy*, Ithaca, N.Y.: Cornell University Press, 2008.

Cheung, Tai Ming, ed., *Forging China's Military Might: A New Framework for Assessing Innovation*, Baltimore, Md.: Johns Hopkins University, 2013.

Cheung, Tai Ming, "The Role of Foreign Technology in China's Defense Research, Development, and Acquisition Process," *Study of Innovation and Technology in China Policy Brief*, January 5, 2014.

Cheung, Tai Ming, Thomas Mahnken, Deborah Seligsohn, Kevin Pollpeter, Eric Anderson, and Fan Yang, *Planning for Innovation: Understanding China's Plans for Technical, Energy, Industrial, and Defense Development*, Washington, D.C.: U.S.–China Economic and Security Review Commission, 2016, pp. 19–24.

"Chinese Quantum Satellite Sends 'Unbreakable' Code," Reuters, August 10, 2017.

Chubb, Andrew, "Propaganda, Not Policy: Explaining the PLA's 'Hawkish Faction' (Part One)," *Jamestown Foundation China Brief*, Vol. 13, No. 15, July 25, 2013.

Chubb, Andrew, "Propaganda as Policy? Explaining the PLA's 'Hawkish Faction' (Part Two)," *Jamestown Foundation China Brief*, Vol. 13, No. 16, August 9, 2013.

Cliff, Roger, "PLA Threat Assessments, Equipment Modernization, Doctrine and Training: 1993–2013," conference paper for CAPS-RAND-NDU International Conference on PLA Affairs, 2013.

———, *China's Military Power: Assessing Current and Future Capabilities*, Cambridge, UK: Cambridge University Press, 2015.

Cliff, Roger, John F. Fei, Jeff Hagen, Elizabeth Hague, Eric Heginbotham, and John Stillion, *Shaking the Heavens and Splitting the Earth: Chinese Air Power Employment Concepts in the 21st Century*, Santa Monica, Calif.: RAND Corporation, MG-915-AF, 2011. As of June 21, 2018: https://www.rand.org/pubs/monographs/MG915.html

Cliff, Roger, Chad J. R. Ohlandt, and David Yang, *Ready for Takeoff: China's Advancing Aerospace Industry*, Santa Monica, Calif.: RAND Corporation, MG-1100-UCESRC, 2011. As of June 21, 2018:
https://www.rand.org/pubs/monographs/MG1100.html

Costello, John, "China Finally Centralizes Its Space, Cyber, and Information Warfare Forces," *The Diplomat*, January 20, 2016.

———, "The Strategic Support Force: China's Information Warfare Service," *Jamestown Foundation China Brief*, Vol. 16, No. 13, February 8, 2016.

Crane, Keith, Jill Luoto, Scott W. Harold, David Yang, Samuel K. Berkowitz, and Xiao Wang, *The Effectiveness of China's Industrial Policies in Commercial Aviation Manufacturing*, Santa Monica, Calif.: RAND Corporation, RR-245, 2014. As of June 21, 2018:
https://www.rand.org/pubs/research_reports/RR245.html

Dai Jinyu, Dong Changjun [戴金宇，董长军], "Air-Space Warfare Theory" [空天一体作战理论] in China Air Force Encyclopedia Editorial Advisory Board [中国空军百科全书编审委员会], *Chinese Air Force Encyclopedia*, Vol. 1 [中国空军百科全书，上卷], Beijing: Aviation Industry Publishing House [北京：航空工业出版社], November 2005, pp. 46–47.

"Drone Joins Four Chinese Ships in Latest Senkaku Intrusion," *Kyodo*, May 18, 2017.

Erickson, Andrew, and Gabe Collins, "Chinese Aircraft Carrier Style! Assessing the First Takeoff and Landing," *Wall Street Journal*, November 27, 2012.

Feigenbaum, Evan A., *China's Techno-Warriors: National Security and Strategic Competition from the Nuclear to the Information Age*, Stanford, Calif.: Stanford University Press, 2003.

"Former Chinese Commander Warns of War if Japan Shoots Down Drone," *Bloomberg*, November 4, 2013.

Garafola, Christina L., "Will the PLA Reforms Succeed?" *China Analysis—European Council on Foreign Relations*, March 30, 2016.

Gertz, Bill, "Top Gun Takeover: Stolen F-35 Secrets Showing Up in China's Stealth Fighter," *Washington Free Beacon*, March 13, 2014.

Goldstein, Lyle J., "How China Views America's Moves in the Asia-Pacific: Worse than Containment," *National Interest*, October 29, 2014.

Griffiths, James, "Beijing Plans Underwater Observation System in South China Sea," CNN, May 30, 2017.

Gunness, Kristen, "China's Military Diplomacy in an Era of Change," conference paper prepared for National Defense University's "China's Global Activism: Implications for US Security Interests," June 20, 2006.

Hallion, Richard P., Roger Cliff, and Phillip C. Saunders, *The Chinese Air Force: Evolving Concepts, Roles, and Capabilities*, Washington, D.C.: National Defense University Press, 2013.

Hannas, William C., James Mulvenon, and Anna B. Puglisi, *Chinese Industrial Espionage: Technology Acquisition and Military Modernization*, New York, N.Y.: Routledge Press, 2013.

Headquarters, Department of the Air Force, *Global Vigilance, Global Reach, Global Power for America*, Washington, D.C., 2013.

Heath, Timothy R., Kristen Gunness, and Cortez A. Cooper, III, *The PLA and China's Rejuvenation: National Security and Military Strategies, Deterrence Concepts, and Combat Capabilities*, Santa Monica, Calif.: RAND Corporation, RR-1402-OSD, 2016. As of June 21, 2018:
https://www.rand.org/pubs/research_reports/RR1402.html

Heginbotham, Eric, Michael Nixon, Forrest E. Morgan, Jacob L. Heim, Jeff Hagen, Sheng Li, Jeffrey Engstrom, Martin C. Libicki, Paul DeLuca, David A. Shlapak, David R. Frelinger, Burgess Laird, Kyle Brady, and Lyle J. Morris, *The U.S.-China Military Scorecard: Forces, Geography, and the Evolving Balance of Power, 1996–2017*, Santa Monica, Calif.: RAND Corporation, RR-392-AF, 2015. As of June 21, 2018:
https://www.rand.org/pubs/research_reports/RR392.html

Holmes, James, and Toshi Yoshihara, "Underestimating China," *The Diplomat*, January 17, 2011.

"Integrated Air-Space Operations (Part 1)" [空天一体作战（上）], *Xinhua* [新华], November 23, 2015.

Jae Ho Chung, "China's Evolving View of the Korean-American Alliance, 1953–2012," *Journal of Contemporary China*, Vol. 23, No. 87, 2014, pp. 425–442.

Johnson, Jesse, "Chinese Air Force Conducts 'Several' Long-Range Drills Near Japan as Military Tells Tokyo to 'Get Used to It,'" *Japan Times*, July 16, 2017.

Kania, Elsa, "China's Strategic Support Force: A Force for Innovation?" *The Diplomat*, February 18, 2017.

Lague, David, "Special Report: China's Hawks Take the Offensive," Reuters, January 16, 2013.

Lewis, John Wilson, and Xue Litai, *China Builds the Bomb*, Stanford, Calif.: Stanford University Press, 1988.

Li, Cheng, and Scott W. Harold, "China's New Military Elite," *China Security*, Vol. 3, No. 4, Fall 2007, pp. 62–89.

Li Gangshe [李刚摄], "Expert: Strategic Support Force Will Link the Entire Operational Process, Be Key to Victory" [专家：战略支援部队将贯穿作战全过程 是致胜关键], *People's Daily Online* [人民网], January 5, 2016.

Liff, Adam P., "China and the U.S. Alliance System," *China Quarterly*, April 24, 2017, pp. 1–29.

Liu Mingfu,"The World Is Too Important to be Left to America," *The Atlantic*, June 4, 2015.

"A Look at China's SAM Shelters in the Spratlys," *CSIS Asia Maritime Transparency Initiative*, February 23, 2017. As of February 23, 2017: https://amti.csis.org/chinas-sam-shelters-spratlys

Mann, James, "U.S. Says Israel Gave Combat Jet Plans to China," *Los Angeles Times*, December 28, 1994.

Mastro, Oriana Skylar, "Why Chinese Assertiveness Is Here to Stay," *Washington Quarterly*, Vol. 37, No. 4, 2014, pp. 151–170.

McGarry, Brendan, "China's Fighters, Drone Look Like U.S. Aircraft," *DefenseTech*, June 20, 2013.

McMillan, Stuart, "Europe's Arms Trade with China," *ASPI Strategist*, February 11, 2014.

Medeiros, Evan S., Roger Cliff, Keith Crane, and James Mulvenon, *A New Direction for China's Defense Industry*, Santa Monica, Calif.: RAND Corporation, MG-334-AF, 2005. As of June 21, 2018: https://www.rand.org/pubs/monographs/MG334.html

Morris, Lyle J., "China's Air Force Freeing Fighters from Scripted Tactics," *Aviation Week & Space Technology*, October 14, 2016.

Morris, Lyle J., and Eric Heginbotham, *From Theory to Practice: People's Liberation Army Air Force Aviation Training at the Operational Unit*, Santa Monica, Calif.: RAND Corporation, RR-1415-AF, 2016. As of June 21, 2018: https://www.rand.org/pubs/research_reports/RR1415.html

Morris, Lyle J., and Eric Heginbotham, "China's PLAAF Pilot Training Program Undergoes Major Overhaul," *National Interest*, October 27, 2016.

Mulvenon, James, "China's 'Goldwater-Nichols': The Long-Awaited PLA Reorganization Has Finally Arrived," *China Leadership Monitor*, No. 49, pp. 1–6.

Mulvenon, James C., Rebecca Samm Tyroler-Cooper, *China's Defense Industry on the Path of Reform*, Washington, D.C.: The U.S.–China Economic and Security Review Commission, 2009.

Mulvenon, James C., Murray Scot Tanner, Michael S. Chase, David Frelinger, David C. Gompert, Martin C. Libicki, and Kevin L. Pollpeter, *Chinese Responses to U.S. Military Transformation and Implications for the Department of Defense*, MG-340-OSD, Santa Monica, Calif.: RAND Corporation, 2006. As of June 21, 2018: https://www.rand.org/pubs/monographs/MG340.html

Nixon, Allen, "China's Growing Arms Sales to Latin America," *The Diplomat*, August 24, 2016.

Ohlandt, Chad, *Chinese Investment in U.S. Aviation*, Santa Monica, Calif.: RAND Corporation, RR-1755-USCC, 2017. As of June 21, 2018: https://www.rand.org/pubs/research_reports/RR1755.html

Panda, Ankit, "Japan to Shoot Down Foreign Drones," *The Diplomat*, October 22, 2013.

Perlez, Jane, "Strident Video by Chinese Military Casts U.S. as Menace," *New York Times*, October 31, 2013.

Pollpeter, Kevin, Eric Anderson, Jordan Wilson, Fan Yang, *China Dream, Space Dream: China's Progress in Space Technologies and Implications for the United States*, Washington, D.C.: U.S.–China Economic and Security Review Commission, 2015.

Ponomarenko, Will, "China Sees Ukraine as Alternative to Russia in Arms Trade, Expert Believes," *Kyiv Post*, May 14, 2017.

Roblin, Sebastien, "China Stole This Fighter from Russia—and It's Coming to the South China Sea," *National Interest*, July 24, 2016.

Saunders, Phillip C., and Joel Wuthnow, "China's Goldwater–Nichols? Assessing PLA Organizational Reforms," *Joint Forces Quarterly*, Vol. 82, 3rd Quarter, July 2016, pp. 68–75.

Saunders, Phillip C., and Joshua K. Wiseman, *Buy, Build, or Steal: China's Quest for Advanced Military Aviation Technologies—Institute for National Strategic Studies China Strategic Perspectives*, No. 4, Washington, D.C.: National Defense University Press, 2011.

Schroen, Gary, *First In: An Insider's Account of How the CIA Spearheaded the War on Terror in Afghanistan,* New York, N.Y.: Presidio Press, 2005.

Scobell, Andrew, David Lai, and Roy Kamphausen, eds., *Chinese Lessons from Other People's Wars*, Carlisle, Pa.: U.S. Army War College Press, 2011, pp. 4–8.

Scobell, Andrew, and Scott W. Harold, "An Assertive China? Insights from Interviews," *Asian Security*, Vol. 9, No. 2, 2013, pp. 111–131.

Shambaugh, David, "Coping with a Conflicted China," *Washington Quarterly*, Vol. 34, No. 1, Winter 2011, pp. 7–27.

Shim, Elizabeth, "China Deploys Air Force on Disputed South China Sea Island," United Press International, October 20, 2016.

Shlapak, David A., David T. Orletsky, and Barry A. Wilson, *Dire Strait? Military Aspects of the China-Taiwan Confrontation and Options for U. S. Policy*, Santa Monica, Calif.: RAND Corporation, MR-1217-SRF, 2000. As of June 21, 2018: https://www.rand.org/pubs/monograph_reports/MR1217.html

Shlapak, David A., David T. Orletsky, Toy I. Reid, Murray Scot Tanner, and Barry Wilson, *A Question of Balance: Political Context and Military Aspects of the China-Taiwan Dispute*, Santa Monica, Calif.: RAND Corporation, MG-888-SRF, 2009. As of June 21, 2018: https://www.rand.org/pubs/monographs/MG888.html

Shou Xiaosong [寿晓松], ed., *Science of Strategy* [战略学], Beijing: Military Science Publishing House [军事科学出版社], 2013.

Stockholm International Peace Research Institute, SIPRI Arms Transfer database. As of July 3, 2018:
https://www.sipri.org/databases/armstransfers

Solem, Erika, and Karen Montague, "Updated—Chinese Hypersonic Weapons Development," *Jamestown Foundation China Brief*, Vol. 16, No. 7, April 21, 2016.

Stefan-Gady, Franz, "China Scores Biggest Military Export Order for Killer Drones," *The Diplomat*, March 2, 2017.

"Taiwan Says Chinese Aircraft Fly Around Island in Weekend of Drills," Reuters, August 13, 2017.

"Telling the Air Force's Story Well Is Another Fist for Dealing with 'Air Noise' and 'Sea Noise'" [“讲好空军故事”是应对“空闹”“海闹”的另一只拳头], *People's Daily Online* [人民网], July 22, 2016.

Tiezzi, Shannon, "3 Goals of China's Military Diplomacy," *The Diplomat*, January 30, 2015.

U.S. Department of Defense, *Annual Report to Congress—Military and Security Developments Involving the People's Republic of China*, Washington, D.C., 2013–2017.

Wang Xiaohong [王晓红], "The Starting Point for Introduce, Digest, Absorb and Re-Innovate" [引进消化吸收再创新的出发点], *People's Daily Online* [人民网], April 3, 2006.

Weisgerber, Marcus, "China's Copycat Jet Raises Questions About F-35," *DefenseOne*, September 23, 2015.

Welsh, Mark A., III, "Global Vigilance, Global Reach, Global Power for America: The World's Greatest Air Force—Powered by Airmen, Fueled by Innovation," *Air and Space Power Journal*, Vol. 28, No. 2, March–April 2014, pp. 4–10.

Wezeman, Siemon T., "China, Russia, and the Shifting Landscape of Arms Sales," Stockholm International Peace Research Institute, July 5, 2017.

"What Does China's Air Force Still Need to Learn from Russia? Su-35 Complements Domestic Production Capabilities," (中国空军还有什么向俄罗斯学? Su-35 补足国内产能) *Sina Military News* (新浪军事), January 2, 2017. As of July 16, 2018: http://mil.news.sina.com.cn/2017-01-02/doc-ifxzczsu6558021.shtml

Whittle, Richard, "New China Drone: Looks Like a Reaper, But . . . ," *Breaking Defense*, September 2, 2015.

Wuthnow, Joel, and Phillip C. Saunders, *Chinese Military Reforms in the Era of Xi Jinping: Drivers, Challenges, and Implications*, Washington, D.C.: National Defense University, 2017.

"Xi Jinping: Speed up the Construction of a Powerful People's Air Force with Integrated Air-Space and Offensive-Defensive Capabilities" [习近平：加快建设一支空天一体，攻防兼备的强大人民空军], *Xinhua* [新华], April 14, 2014.

"Xi to Head Civil-Military Integration Body," *Global Times*, January 23, 2017.

Yan Xuetong, "From 'Keeping a Low Profile' to 'Striving for Achievement,'" *Chinese Journal of International Politics,* Vol. 7, No. 2, Summer 2014, pp. 153–184.

Yasuhiro, Matsuda, "An Essay on China's Military Diplomacy: Examination of Intentions in Foreign Strategy," *National Institute of Defense Studies Security Reports*, No. 7, December 2006, pp. 1–40.

Yeo, Mike, "China Deploys New Anti-Submarine Aircraft to Fringes of South China Sea," *Defense News*, June 22, 2017.